Steam Traction Engineering

VFK
Eiching

**STEAM TRACTION
ENGINEERING**

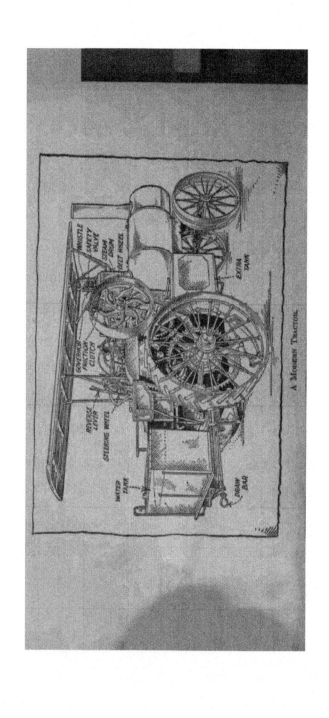

A Modern Tractor.

STEAM TRACTION ENGINEERING

A BOOK FOR OPERATING ENGINEERS

BY
S. R. EIGHINGER
AND
MANCIUS SMEDES HUTTON

ILLUSTRATED

D. APPLETON AND COMPANY
NEW YORK AND LONDON
1916

Copyright, 1916, by
D. APPLETON AND COMPANY

Printed in the United States of America

PREFACE

The purpose of this book is to present in as compact a form as possible those principles which underlie a thorough knowledge of steam traction, farm or portable engines.

It has been the object of the authors to deal with the subject herein treated in a practical way as applied to everyday work rather than to discuss theories and mathematical problems. The book has been written in simple language and in such detail that an engineer of average intelligence will have no difficulty in following the instructions given.

The name, nature and function of the principal parts of the steam boiler and engine of these machines have therefore been set forth with the directions for their operation or use. Considerable space has been given to troubles which may arise in the course of operation, to the method to be followed to determine their causes and how they may be removed or the part affected repaired.

In order to help the engineer to follow the functions and operations of the various parts of the boilers and engines the authors have illustrated the book with a large number of sectional views.

It is hoped that many operators of engines will profit by the facts given in this book and will therefore be better able to cope with the many trying problems that will be met from day to day.

New York.

S. R. EIGHINGER.
MANCIUS S. HUTTON.

CONTENTS

CHAPTER		PAGE
	BOILERS AND BOILER ACCESSORIES	1

Vertical Boilers—Return Flue Boilers—Locomotive Boilers—Glass Water Gauge—Gauge Cocks—Steam Gauge—Safety Valve—Blow-off Valve and Pipe—Fusible Plug—Blower—Boiler Feed Pumps—Crosshead Pump—Geared Pump—Clark Pump—Marsh Pump—Injector—Feed Pipes and Their Fittings—Valves.

I. CARE AND MANAGEMENT OF BOILERS 41

Starting a New Boiler—Firing with Coal—Firing with Wood—Firing with Straw—Grates—Tubes—Cleaning a Boiler—Foaming and Priming—Scale—Low Water—Boiler Explosions—Preparing a Boiler for Storage.

II. THE REPAIRING OF BOILERS 68

Tools Required—Putting on New Tubes or Flues—Replacing Staybolts—Tightening Leaky Tubes—Repairing Leaks at Staybolts—Leaking Seams—Applying a Patch—Rivet Leaks—Cutting a Hand Hole—Fitting Gauge Glasses—Grinding Leaky Check Valves—Grinding Globe and Angle Valves—Grinding Stop Cock—Fitting Throttle and Gate Valves—Grinding a Safety Valve—Testing and Adjusting Steam Gauges—Testing Boilers.

V. THE ENGINE MECHANISM 103

Classes of Engines—The Cylinder and Steam Chest—Engine Frame—Crosshead and Guides—Piston Rod to Crosshead—Connecting Rod and Boxes—Crank

CONTENTS

CHAPTER   PAGE

Shaft—Main Bearing—Flywheel and Friction Clutch—Pin Clutch—Throttle—Cylinder Cocks—Displacement Lubricators—Hydrostatic Lubricators—Oil Pumps—Oil and Grease Cups—Governor—Engine Mountings—Traction Gearing—Compensating Gear—Starting an Engine.

V. CARE AND MANAGEMENT OF ENGINES 149

Starting an Engine—Hot Boxes—Lubrication of Gears—Lubrication—Knocks and Pounds—Keeping Gears in Line—Counter and Idler Shafts—Rear Axles—Front Axles—Steering Attachment—Water Tanks—Tender—Storage of an Engine—Handling a Traction Engine.

VI. VALVE GEARS AND VALVE SETTING 174

Lap—Lead—Expansion—Direct and Indirect Valves—The Plain Piston Valve—The Allen Valve—The Double-ported Balanced Valve—The Giddings Valve—The Woolf Valve—Other Compound Engine Valves—The Gould Balanced Valve—The Newton Balanced Valve—The Non-Leak Balanced Valve—The Baker Balanced Valve—Valve Gears—Directions for Setting the Plain Eccentric and a D Valve—Reversing Valve Gear—The Link Reverse—The Woolf Reversing Gear—The Springer Reverse Gear—The Marsh Valve Gear—The Arnold Reverse Gear—The Reeves Reverse Gear—Other Styles of Valve Gears.

VII. REPAIRING OF ENGINES 225

Adjusting Crank and Crosshead Pin Bearings—Adjusting Main Bearings—Adjusting the Crosshead—Adjusting Eccentrics—Fitting a New Crosshead or Slides—Fitting Crosshead Pin and Brasses—Fitting Crank Pins and Brasses—Cutting Oil Grooves in Crank and Crosshead Pin Brasses—Babbitting Boxes—Repairing Governors—Fitting a New Governor Stem—Repairing the Friction Clutch—Fitting Cut-off Valves—Fitting Piston Rings and Cylinders—

CONTENTS

CHAPTER   PAGE

Packing an Engine—Fitting Gaskets—Packing Piston and Valve Rods.

APPENDIX 279

Boiler Arithmetic—Steam Engine Arithmetic—The Pull on the Drive Belt—Examination Questions and Answers.

GLOSSARY OF TECHNICAL TERMS 303

INDEX 313

LIST OF ILLUSTRATIONS

A modern tractor . . . . . . *Frontispiece*

FIG.	PAGE
1.—Standard vertical boiler	2
2.—Submerged tube vertical boiler	3
3.—Return tube boiler	4
4.—Water front return tube boiler	4
5.—Water bottom locomotive boiler	5
6.—Open bottom locomotive boiler	6
7.—Glass water gauge	7
8.—Gauge cock	9
9.—Section of a single spring steam gauge	11
10.—Steam gauge siphon	12
11.—Section of pop safety valve	13
12.—Fusible plug	15
13.—Plunger force pump	18
14.—Crosshead pump	19
15.—Geared pump	22
16.—Clark pump	23
17.—Marsh pump	25
18.—Automatic injector	32
19.—Diagrammatic view of boiler scale in tubes	57
20.—Diagrammatic view of scum on surface of water	57
21.—Diagrammatic view of scale accumulation on crown sheet and staybolts	58
22.—Diagrammatic view of cracks around rivets and staybolts	59
23.—Diagrammatic view of scale on tube ends	59
24.—Compound feeder	61
25.—Boiler repair tools	68
26.—Cracks around staybolts	77
27.—Plug inserted to stop staybolt crack	78
28.—Applying a patch to a main flue of a boiler	80
29.—Correct method of holding a calking tool	81
30.—Applying a patch on the fire-box side of a boiler	82
31.—Heel bar	85

LIST OF ILLUSTRATIONS

FIG.		PAGE
32.—Use of a heel bar		85
33.—View showing method of adjustment of steam gauge		95
34.—Enlarged view of pinion and segment of steam gauge		98
35.—Testing pump		100
36.—Engine cylinder and valve		104
37.—Engine box frame with locomotive guides		107
38.—Girder type engine frame with Corliss crosshead and guides		107
39.—Top view locomotive style crosshead and guides		108
40.—Ordinary Corliss crosshead		109
41.—Improved Corliss crosshead		110
42.—Connecting rods		112
43.—Crank shafts		115
44.—Main shaft boxes		116
45.—Eccentric		117
46.—Flywheel and friction clutch		118
47.—Pin clutch		121
48.—Throttles		122
49.—Water displacement lubricator		125
50.—Sectional view hydrostatic lubricator		126
51.—Double connection lubricator		127
52.—Single connection lubricator		128
53.—Method of attaching double connection lubricator to one opening in steam pipe		129
54.—Plain oil cup		133
55.—Glass oil cups		134
56.—Sectional view of spring compression grease cup		135
57.—Governor		137
58.—Independent under-mounted mounting		142
59.—Independent top-mounted mounting		143
60.—Top-mounted side-hung engine mounting		143
61.—Top-mounted rear-hung engine mounting		144
62.—Transmission gearing		145
63.—Compensating gear		147
64.—Worn steering gear		163
65.—Screw steering gear		164
66.—Two-wheeled tender truck		167
67.—Plain D valve		175
68.—Double-ported piston valve		179
69.—Plain piston valve		181
70.—Allen valve		182
71.—Double-ported balanced valve		183

LIST OF ILLUSTRATIONS

FIG.		PAGE
72.—Giddings valve		184
73.—Woolf valve		186
74.—Woolf valve. Face view		187
75.—Gould balanced valve		189
76.—Newton balanced valve		190
77.—Non-leak balanced valve		191
78.—Baker balanced valve		192
79.—Plain eccentric valve gear		193
80.—Centering an engine		195
81.—Link reverse		198
82.—Woolf reverse		204
83.—Springer curved block reverse gear		209
84.—Marsh reverse		210
85.—Arnold reverse		215
86.—Reeves new reverse and expanding gear		219
87.—Grooving brasses		245
88.—Lining main shaft		247
89.—Grooving main shaft boxes		250
90.—Hollowing out brasses for babbitting		254
91.—Crosshead shoe oil grooves		259
92.—Babbitting cannon bearings		262

INTRODUCTION

The evolution of the steam traction engine may be divided into four rather distinct stages.

1. The original form was a stationary type in which the engine was mounted entirely independently of the boiler and on a separate foundation. This style was found to be very unsatisfactory for such work as threshing, running small saw mills and other portable farm machinery due to the trouble and inconvenience of moving and resetting the engine.

2. This caused designers to build engines in which the combination was self-contained and mounted on skids or truck wheels. The steam engine was placed on top of the boiler.

3. The next stage was the designing of an engine which could propel itself. This was accomplished first by the addition of a sprocket wheel and chain which was replaced by a train of gears in the later designs. This converted the portable engine into a true traction one. The engine was also further improved by the addition of a steering gear.

4. During this stage of development the engine was increased in strength and power so that it not only could pull itself but would be strong enough to pull plows, move houses and do heavy hauling. This made the engine deliver power not only at the flywheel but also at the draw bar.

Notice the difference between a locomotive, a stationary engine and a traction engine. A locomotive moves at a high rate of speed over smooth rails and delivers power

INTRODUCTION

only at the draw bar. The stationary engine does not move and delivers its power only at the flywheel. The traction engine moves slowly over good and bad roads and across fields to the place where it is needed and it can deliver power both at the flywheel and the draw bar according to the type of power required.

A steam tractor consists essentially of three parts: (1) the boiler in which the steam is formed; (2) the steam engine in which the steam is made to produce motion and useful work, and (3) other mechanisms necessary in order to control the machine, i.e., steering attachment, compensating gear, etc. The steam tractor of today has a boiler mounted on wheels. The steam engine is located on the top of the boiler, usually at the front end.

The steam which is generated by heating water in a steel shell (the boiler) closed to the atmosphere, passes through a pipe to the steam chest of the engine valve located over the cylinder. The reciprocating motion given to this valve allows a certain quantity of the steam to pass alternately to each side of the piston in the cylinder. The steam, due to its inherent tendency to expand, causes the piston to move alternately back and forth in the cylinder according to the side on which the steam is expanding. This reciprocating motion of the piston is communicated to the crank through the operation of the piston rod, crosshead and connecting rod. The crank, which is attached at one end to the connecting rod and at the other to the shaft of the flywheel, due to its construction imparts a rotating motion to the latter. The tractor is propelled by means of a chain of gears from this shaft to the two rear wheels.

It is well to remember the following properties of steam: (1) that the temperature of steam in contact with water depends upon the pressure under which it is generated and that as the pressure is increased, as by the steam being generated and confined in a closed vessel, its temperature and

INTRODUCTION

that of the water in its presence increases. The steam tables give the corresponding temperature to each increase in pressure. (2) During the time that the water is changing from the liquid to the gaseous (steam) state there will be no increase in temperature. (3) If the steam is at a higher temperature than that due to its pressure it has become superheated. Water is to be found in one of three states: solid (ice), liquid, and gaseous. When it is passing from the liquid to the solid form it expands in volume. For this reason no water should be left in the pipes and steam cylinder in cold weather.

STEAM TRACTION ENGINEERING

CHAPTER I

BOILERS AND BOILER ACCESSORIES

Boilers can be divided into two classes. In the first class the furnace is a part of the boiler. In the second it is a separate unit. Those of the first class are known as internally fired boilers while the others are known as externally fired. In farm, traction, and portable work it is necessary that the boiler be of the first class as it is impracticable to have the furnace built of masonry outside of the boiler.

Each of these classes is subdivided into two types. In one the boiler contains fire tubes and in the other water tubes. In the fire-tube boilers the products of combustion pass through the tubes, and the water surrounds them, while in the other type the process is reversed; i.e., the gases pass around the tubes which contain the water. In water-tube boilers the tubes are inclined at an angle in order to give better circulation to the water, which makes this type larger and heavier than the other. The fire tubes are always perfectly horizontal or vertical.

For these reasons the boilers used on farm, traction, and portable machines are of the internally fired fire-tube group. This will reduce the study of boilers to three types under this one group, namely, the vertical, return flue, and the locomotive boilers.

STEAM TRACTION ENGINEERING

tical Boilers.—This type illustrated in Fig. 1 has a
rical shell set on end with the grate at the bottom and
e tubes running vertically to the flue or stack. The
x is smaller than the shell and is connected to it at the
either by a wrought-iron ring or by a flanged plate,

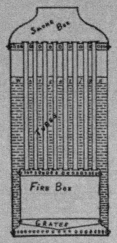

FIG. 1.—STANDARD VERTICAL BOILER.

llowing the water to circulate completely around the
xx, which is fastened to the outer shell by means of
olts. Should the upper flue sheet be lowered ten to
inches so that the lower part of the smoke flue is
the water line, the boiler is known as the submerged
ype (Fig. 2).

vertical type of boiler has the advantage that it occu-
ery little space and is a rapid and easy steamer. It is,

BOILERS AND BOILER ACCESSORIES

however, open to the following objections: (1) It carries a relatively small amount of water and contains a very small steam space for its nominal horsepower. (2) It foams or primes very easily. (3) It is hard to keep the upper ends of the flues tight when the boiler is working heavily because they are not surrounded by water. (4) It requires

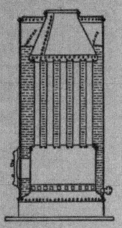

FIG. 2.—SUBMERGED TUBE VERTICAL BOILER.

careful attention on the part of the engineer to prevent the pressure from passing from the safe to the danger point because they hold only a small amount of water. This type is being superseded by the other two although there are a few still being built.

Return Flue Boilers.—This type approaches in principle the Scotch marine boiler. It has a cylindrical shell resembling a large drum in shape, which is set horizontally. In

STEAM TRACTION ENGINEERING

the lower half of the boiler there is a cylindrical flue containing the furnace which terminates in a back chamber at

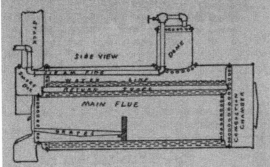

Fig. 3.—Return Tube Boiler.

the rear. Around the sides and over this furnace are a large number of fire tubes extending from the back connec-

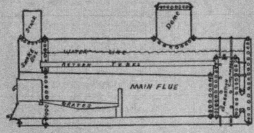

Fig. 4.—Water Front Return Tube Boiler.

tion to the uptake to the smokestack at the front (Fig. 3). The flue may have either a corrugated or flat surface. The

BOILERS AND BOILER ACCESSORIES

corrugation increases enormously the resistance to deformation by pressure, but it has the drawback that it is difficult to keep it clean. Some makes of boilers of this type have water around the back connection as well as between the lower part of the flue and the shell (Fig. 4). Back connections which are not protected by means of water have cast-iron plates to prevent burning out of the shell. The hot gases pass back through the flue to the back connection

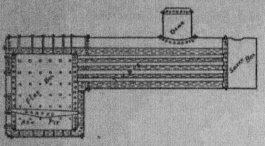

FIG. 5.—WATER BOTTOM LOCOMOTIVE BOILER.

and forward again through the tubes to the smokestack at the front.

The advantages of these boilers for farm engines are: (1) They are self-contained and portable. (2) They are steady steamers, maintaining a uniform pressure under varying loads and conditions, owing to the large volume of water which acts as a reservoir of heat. (3) They are quickly and easily accessible for cleaning and making repairs. (4) They require few, if any, stay bolts and are considered of safe construction. (5) They will burn the lowest grade of fuel and when properly fired are economical in amount of fuel required, as well as being practically smokeless.

They are open to the objection of having a very limited grate area per horsepower and when used in connection with farm machinery their necessary short length is often a disadvantage.

Locomotive Boilers.—This type consists of a rectangular fire box and a cylindrical shell. Numerous fire tubes are contained within this shell and extend from the fire box to

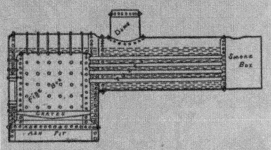

FIG. 6.—OPEN BOTTOM LOCOMOTIVE BOILER.

the smoke box; to the latter the stack is connected (Fig. 5). The fire box is smaller than the outer casing to allow water to circulate freely around this portion of the boiler. In some designs the water is allowed to circulate around the bottom of the furnace as well, space being provided for the same. This design is known as a water-bottom boiler to distinguish it from those which do not have this provision and are provided instead with a ring (Fig. 6).

The advantages of this type of boiler are: (1) They are self-contained, requiring no masonry and therefore very advantageous for use in traction and portable engine construction. (2) They are quick steamers, easy to handle, and are provided with plenty of heating surface. (3) The

BOILERS AND BOILER ACCESSORIES

fire is entirely surrounded by water, and this makes them economical in the use of fuel.

The great disadvantage of this type is that they require many stays to hold the flat plates around the fire box in place against the external pressure which tends to collapse them. Also the trouble due to unequal expansion of the

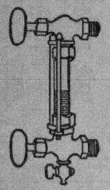

FIG. 7.—GLASS WATER GAUGE.

inner and outer plates has to be considered. In the newer constructions these plates are corrugated, thus increasing their strength and also avoiding much of the trouble due to failure of the stay bolts. The screwed stay bolts pass through both plates and are riveted. The flat crown sheet or upper plate of the fire box is held up against collapse either by radial bolts or stays attached to the outer shell.

In this construction of boiler the products of combustion only give up their heat to the water when passing through the tubes from the fire box to the stack, while in the return flue type these products give up their heat through a

STEAM TRACTION ENGINEERING

longer period of time in going through the flue and returning through the tubes to the stack.

In both the return tubular and locomotive type of boiler a large steam drum should be provided and located in such a position that the required amount of dry steam can be readily obtained for the engine at all times, especially if it is going up or down hill.

All boilers are fitted with several washout plugs and hand holes, the number varying with the size and type.

Glass Water Gauge.—This is an instrument to tell at a glance the quantity of water in a boiler. It consists of a glass tube about a foot long with its lowest end two inches above the highest part of the crown sheet and in connection with the water space of the boiler. Its upper end is connected with the steam space (Fig. 7). The gauge glass may be connected directly to the shell or to a special fixture called the water column whose two ends are connected to the steam and water space as above specified. The tube is fitted at each end to a valve by means of packing nuts and packing. Below the bottom valve is attached a drain cock. The presence of water in the gauge, provided the boiler is level, denotes that all fire surfaces are covered with water, but too much dependence must not be laid on it, as it often happens that sediment and scale choke the connections to the tube, causing it to give a wrong reading. For this reason it is essential that the gauge be blown out at least once a day if not oftener. To do this, close the bottom valve and open the drain cock. This allows steam from the upper valve to blow out the sediment. Now close the upper valve and open the lower, allowing the water to wash out the sediment in the lower connection. After this close the drain cock and slowly open both valves wide, at the same time noticing whether there is any change in the level of the water. For method to be followed when replacing a broken gauge glass see page 88.

BOILERS AND BOILER ACCESSORIES

During the interval when the gauge is out of commission due to a breakage of the glass or for any other reason, the level of the water can be determined approximately by means of the gauge cocks.

Gauge Cocks.—Another way to tell the level of the water in the boiler is by means of the gauge cocks or valves (Fig. 8). There are usually three of them to be found on each boiler. The lower one is placed at the lowest level that the water may safely attain and the uppermost cock at the highest level the water can be safely carried. The

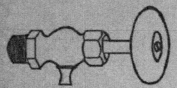

FIG. 8.—GAUGE COCK.

third valve is placed midway between these two extremes or at the level at which the water should be kept. They are placed on the shell or head of the boiler, although they are often attached to the water column. It is just as essential that these cocks should be kept free from sediment as it is for the water gauge. For this reason they should be blown out frequently. Engineers sometimes neglect these cocks and place their faith entirely upon the water gauge, allowing the cocks to become choked with scale and sediment. They are placed on the boiler for the purpose of acting as a check to the readings of the water gauge and as such should be kept in good working condition. An engineer should at all times *know* the *exact level* of the water in the boiler.

To determine the water level by means of the gauge cocks, open them in succession, beginning with the lowest

STEAM TRACTION ENGINEERING

one, and notice out of which one steam or water appears. Water should always appear at the lowest one and steam from the highest, unless there is something radically wrong with the amount of water in the boiler. Either steam or water will appear at the middle cock according to whether the water is above or below this point. At first it will be hard to distinguish between the steam and the water but after a little practice they can be told apart by the difference in color and sound. Steam hisses with a dry sound and is light blue in color close to the cock, while water does not hiss but sputters and will appear milky white.

In the morning before starting the fire, see that the water level is at the proper height and the valves connecting the gauge glass to the boiler are wide open. It is advisable, when there is pressure enough, to blow out at this time all sediment which may have collected during the night in the connections to the water gauge and the gauge cocks. Should the boiler be provided with a water column, this should be blown out.

Steam Gauge.—This is an instrument attached to the boiler for the purpose of indicating the pressure of the steam in the boiler. While there are several different types of gauges, the double-spring one is most common on farm engines. The interior construction of a single spring type is shown in Fig 9. In this design there is a small curved flattened tube B which is closed at its upper end and open at the other A. To the closed end are attached toggle levers b which are in turn fastened to a toothed segment D. This segment is in mesh with a small pinion on which a pointer is attached. The pressure entering at A has a tendency to cause the tube B to straighten itself out, which in turn imparts a movement to the pointer through the medium of the multiplying pinion, segment and toggle levers. The amount of this movement varies according to the pressure of the boiler. The pointer moves over a graduated dial

BOILERS AND BOILER ACCESSORIES

which shows the pressure in pounds per square inch when the spring tube has been correctly calibrated.

This instrument and those mentioned before are among the most important ones belonging to the boiler. They should be kept in good working order as upon them depends the life of the operator and those near the engine. Defec-

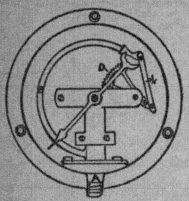

FIG. 9.—SECTION OF A SINGLE SPRING STEAM GAUGE.

tive steam and water gauges cause a great many boiler explosions. There are three parts about one of these gauges which are the common causes of improper working. The first of these is the weakening of the tube or spring in the course of time. Second, the various joints and parts rust. Third, the pointer gets out of place on its spindle. It is a good plan to have the steam gauge tested at least twice a year or whenever there is any doubt as to its accuracy by a qualified outside person having the necessary instrument. *Never* under any circumstances attempt to change a steam gauge to conform with what *in your opinion* is the right pressure at which the safety valve blows off. Again *know* that the gauge is correct before making any adjustment about the safety valve.

On account of the fact that steam will ruin the spring of a gauge, it is necessary that it should be connected to the boiler by means of a siphon or coiled pipe (Fig. 10). It is essential that this connection should be kept free of sediment and full of water. There is usually placed a shut-off

FIG. 10.—STEAM GAUGE SIPHON.

cock between the boiler and the gauge, which should be kept always open. By means of this arrangement the gauge can be unscrewed and the siphon and connections blown out with steam pressure.

The steam and water gauges should be the best that can be procured in the open market.

Safety Valve.—This device, as its name implies, is attached to the boiler to prevent the steam pressure rising above a designated point, which is usually the maximum safe working pressure. Boilers are designed and constructed to withstand a certain fixed maximum steam pressure, and should they exceed this amount they are likely to rupture and explode. The valve found most commonly in use on boilers of farm engines is known as the spring type.

BOILERS AND BOILER ACCESSORIES

The interior construction of one of them is shown in Fig. 11. It consists of a body in which there is a plate or disk fitted over a hole in connection with the steam space of a boiler. This disk C is held against the boiler pressure by means of a helical spring, which is so adjusted that when the steam

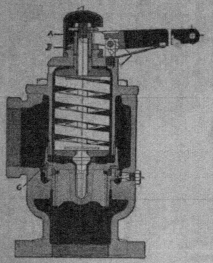

FIG. 11.—SECTION OF POP SAFETY VALVE.

pressure reaches the designated point the disk is raised from its seat, permitting the excess of pressure to escape. The valve can be set to blow off at any desired pressure by loosening or tightening the adjusting screw A on the top of the spring. After the valve is once set, see to it that the adjusting screw is locked by means of the lock nut B. In

STEAM TRACTION ENGINEERING

getting up steam always raise the valve from its seat by pulling down the little lever on its side. Keep it in good working order and at least once a day allow it to blow off.

The safety valve is usually connected directly to the boiler. Where it is necessary to use a nipple it should be made as short as possible. The use of ells should be avoided and there should be no valve between the boiler shell and the safety valve.

One of the worst practices that can be followed is to screw down the safety valve in order to get more power when an engine gets "stuck."

It is sometimes advisable to have two safety valves on each boiler although at the present time it is not universally required by law. With two valves, one is set to blow off at the designated pressure while the other is set for three to five pounds higher.

These valves should be of the extra heavy pattern and of a good standard make.

Blow-off Valve and Pipe.—All boilers should be cleaned of sediment and scale which will collect at the lowest and coolest point. Therefore at this point a pipe is attached to the boiler leading to the outer air. This pipe is known as the blow-off pipe and will have with it as close to the boiler as convenient a blow-off cock or valve. Any good steam cock, gate or globe valve can be used for this purpose but on account of the severe use they are put to they should be of high quality and heavy pattern. The valve should be located at the lowest part of the boiler and in such a position that it can be worked without difficulty and danger of the operator's being scalded in opening and closing it. Generally boilers should be blown down two gauges once or twice a day and entirely emptied once a week, unless the condition of the water renders more frequent emptying necessary. If upon closing the valve it should leak, do not try to force it tight, but instead open it and close it again

BOILERS AND BOILER ACCESSORIES

with a quick turn. This will blow out the mud and scale between the seat and disk. Trying to close a leaking valve by force will result in either springing the valve or cause the seat and disk to be so cut that thereafter it will never be tight.

A short piece of pipe or nipple should be screwed into the end of the valve so that the water will be discharged in the form of a straight jet.

Another kind of blow-off valve sometimes found is known as the "surface blow-off" and is located at about the same

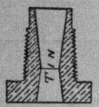

FIG. 12.—FUSIBLE PLUG.

height as the working level of the water in the boiler. Its purpose is to get rid of the scum and foam that rises to the surface of the water when working the boiler with some kinds of water. This blow-off valve should be opened for a few moments at a time several times a day.

Fusible Plug.—One of these plugs is screwed into the shell of a boiler to give warning to the engineer of low water. It consists of a brass shell threaded on the outside with a standard pipe thread. The inside cavity is filled with an alloy which will melt at a heat a little above that of the ordinary steam pressure of the boiler (Fig. 12). As long as the plug is covered by water the fusible alloy is kept from melting by the rapid absorption of the heat by the water, but should the water sink below the level of the plug the heat will melt it. This permits the steam to be discharged, giving a warning of low water and practically emptying the boiler. The location of a fusible plug in the boiler varies with the different types but in general they are placed at the lowest permissible water level in the direct path of the products of combustion and as near the primary combustion chamber as possible. They project out about one inch on the inside of the boiler. The alloy which has been used up to the present time is mainly either lead or some form of babbitt. Pure tin with a melting point of between 400° and 500° F. will probably be the metal used hereafter.

Particular care should be taken to keep them clean as they are apt to get corroded or coated with lime. In this case they cannot be depended upon to do their work. Should this happen the crown sheet will get badly burned or bagged and an explosion may result when the water which is pumped into the boiler comes in contact with the red-hot crown sheet. When examining the boiler the plug should be carefully looked after and if necessary the deposits should be scraped off. Should the metal seem corroded the plug should be replaced with a new one or the old one filled with new metal. When replacing a plug graphite instead of white lead should be used on the threads. These plugs should be of brass having a large square head, so that in removing them a good grip can be taken with a wrench.

Blower.—The blower consists of a small pipe running from a point in the steam space of a boiler to the smokestack. The end of the pipe in the smokestack should point straight up in the middle of it, as otherwise it will lose its effectiveness. The purpose of the pipe is to improve the natural draft of the boiler whenever necessary. The steam controlled by a valve escaping through this pipe will cause a partial vacuum in the smoke box by driving out the air.

BOILERS AND BOILER ACCESSORIES

This will cause a strong draft of fresh air to come through the fire, flue and tubes to fill up the vacuum. Little or nothing can be gained by using the blower with less than fifteen pounds pressure. With less than this the consumption and waste of steam will be greater than the increase in the draft will generate. Only in very exceptional cases should the blower be used when the engine is running and discharging its exhaust through the stack. Its usefulness will be greatly impaired should the end of the nozzle become covered with lime or should it get out of center line in the stack. Should the fires be banked at night and dampers not fit tight, a leaky blower valve will cause enough draft to keep the safety valve singing. Unless the boiler was pumped nearly full of water, either a burnt crown sheet or melted fusible plug will be found in the morning.

Boiler Feed Pumps.—There are three groups of pumps, used in connection with farm engines, divided according to the source of power which operates them. In the first group are those in which the plunger or piston rod is attached directly to the farm engine mechanism, such as in the crosshead or geared pumps. In the second group the pumps are driven by independent engines, either gas, oil or steam, attached directly to them. In the third group the pump is driven by its own steam which is furnished by the boiler. There are other groups of pumps but they are hardly ever found in connection with farm engines and therefore no mention will be made of them.

All of the above groups can be divided into two classes: piston and plunger pumps. A plunger pump is one in which the inside circumference of the pump cylinder is a little larger than the circumference of the long moving cylinder within it, thus allowing a small quantity of water to remain in the cylinder during each stroke. In the piston pump there is no difference in the circumference of the

STEAM TRACTION ENGINEERING

cylinder and the short moving piston. The piston, as in the case of a steam engine, fits snugly around the cylinder walls.

The pumps in the above groups may also be either single cylinder or duplex. The advantages and disadvantages of these various types will be given a little later.

Besides a pump the boiler may be fed with water by means of an injector. It is advisable to have two methods of feeding the boiler, so that in case of the failure of one, the other can be used.

It is of vital importance that the boiler feeder should at all times be in good operating condition for the following two reasons: (1) steam cannot be made without water;

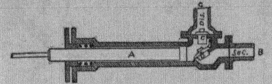

FIG. 13.—PLUNGER FORCE PUMP.

(2) a boiler is a very dangerous piece of machinery unless it is supplied with enough water to keep the heating surfaces always covered with it so as to prevent overheating the metal.

Crosshead Pump.—This is a simple plunger pump (Figs. 13 and 14) consisting of a plunger from five-eighths to one and an eighth inches in diameter, according to the size of the engine, working in a cylinder slightly larger in bore than the plunger. The cylinder is somewhat longer than the stroke of the plunger. It is fitted with a packing gland at the end where the plunger rod comes through the cylinder, while the other end is connected to the valve chamber, which contains a suction and a discharge valve besides an air chamber. The plunger rod is made of steel and extends

BOILERS AND BOILER ACCESSORIES

from the plunger to the pump post on the crosshead of the engine. The plunger travels the full stroke of the cylinder between the packing nut and the valve chamber.

The operation of the pump is as follows: When the plunger A (Fig. 13) is drawn outward it creates a vacuum in the cylinder. The pressure of air on the water in the supply tank (provided it is open to the atmosphere) will force it up the suction pipe through the suction valve B and into that part of the cylinder left vacant by the receding plunger. If the suction pipe is not under water, the

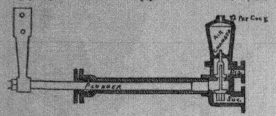

FIG. 14.—CROSSHEAD PUMP.

pump will not work. When the plunger starts on its return stroke, the suction valve is forced shut while the discharge valve C opens; the water passes into the air chamber and pipe leading to the boiler. Between the pump discharge and the boiler there should be a check and stop valve.

Theoretically a pump should lift water thirty-four feet at the sea level, but because of friction in the pipes, leakage of valves and packing and lowered barometric pressure, the practical lift will be reduced down to twenty-four or twenty-six feet.

A pump of this type cannot be started unless the engine is running. In starting be sure first to open the stop valve between the pump and the boiler or something will be sure to burst. Next open the valve on the suction pipe and also

STEAM TRACTION ENGINEERING

the pet cock on top of the air chamber. As soon as water comes out of the pet cock it should be closed. Should the pump not start at this point to perform its functions some part is in need of attention.

One of the best indications that the pump is working properly is the clear sharp click of the valves.

Any one of the following may be the cause of the pump not working.

1. If the pump does not start or pick up water,

a. There may be no water in the tank or not enough to cover the suction pipe;

b. There may be an obstruction under the suction valve which prevents it from closing tight;

c. There may be an obstruction or leak in the discharge check valve. Before the cap is taken off this valve, the stop valve next to the boiler must be closed or the person working will be badly scalded. A way to locate a faulty check valve on the discharge line is to feel whether the pump is hot or cold. A cold pump is a good indication that it is in good working order. Hot water coming out of the pet cock indicates that the check is not seating properly. Should there be plenty of water in the tank, place the hand loosely over the end of the pipe and feel the suction. If there is not any or only a small amount, then it is possible that the trouble is in the discharge valve.

2. If the water is forced in and out again of the tank, either no water or a very small amount going to the boiler,

a. The trouble is due to the suction valve. If something gets under either of these valve seats to prevent their closing tightly, the water which under normal circumstances would be forced by the plunger out through the discharge valve will instead return to the tank.

If upon examining the pump valves, they are found rough or burred they can be ground smooth without much trouble. To do this, bore a hole about an inch deep in a

BOILERS AND BOILER ACCESSORIES

block of wood large enough to admit the valve. Fill the hole nearly full of emery dust wet with oil. Place the valve in the hole and with a brace turn the valve around quickly a few minutes. In place of emery dust, fine grindstone grit or fine sand can be used.

3. If the pump starts all right but fails to pump the proper amount of water, look at the following points for the possible cause:

a. Leaky or faulty plunger packing. It may be remedied by giving the stuffing nut a turn or two. If this does not obviate the trouble it should be repacked at the first opportunity. Use a good spiral packing cut in rings, braided asbestos or hemp soaked in oil. (See page 276.)

b. Leaky suction pipe. To test this pipe, plug the lower end tightly and apply soapsuds to the hose and pipe joints. The air will locate the leaks. Another cause is a loose lining in the suction hose. The defective part should be cut out or if necessary a new hose should be inserted.

c. Leaky globe and gate valves on the pipe lines. Instructions for regrinding these valves are given on page 91.

The nipples where the water enters the boiler may become clogged with lime until the hole is so small that enough water cannot be forced through.

Considerable trouble with the pump can be eliminated by having a good strainer at the bottom of the suction hose. Every once in a while the strainer should be cleaned, especially if the water is dirty.

Obstructions under the valves are sometimes hard to find and may be easily overlooked. A piece of wood may get between the valve and its seat. When the valve is taken down for examination the piece will drop back into the pipe where it is lost to view. In such a case it would be necessary to remove the pipe.

Care should be taken to keep the cylinder in line with the plunger and the latter well secured to the pump post

STEAM TRACTION ENGINEERING

and crosshead. The crosshead should be kept adjusted so that it does not have any play in the guides. If they get out of line they will cut the packing, rod and gland so that it will be impossible to keep the pump properly packed, besides putting a heavy strain on the plunger which may cause it to break.

It sometimes happens that the suction valve is made with a stem that fits into a hole in the bottom of the discharge valve. If the stem should be slightly bent, the discharge valve will not be able to seat squarely.

Geared Pump.—This pump is identical with the cross-

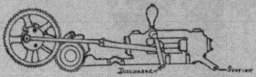

FIG. 15.—GEARED PUMP.

head one mentioned above, except that it is operated by a crank shaft which is joined to the main shaft of the engine by means of two or more gears. A connecting rod from this crank shaft gives motion to the plunger. For the above reason the pump can be used only when the main engine is running. The pump is built either single or double acting and with one or several cylinders. Fig. 15 illustrates a double acting one in which there are two cylinders having a common plunger. Each cylinder has its own valve chamber and valves, but they are connected to a common suction pipe, discharge pipe and air chamber. This pump works on exactly the same principle as the crosshead one except that while one cylinder is forcing water through the discharge valve the other is taking in water through the suction valve. This results in a more steady stream of

BOILERS AND BOILER ACCESSORIES

water and the piping is subject to a less severe strain than is the case in the single-acting pump. The same result of smooth continuous acting with this type of pump can be

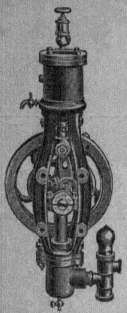

FIG. 16.—CLARK PUMP.

obtained by having two single-acting plungers side by side and driven by two cranks on the crank shaft. Here it is more usual to control the amount of water by a valve on the suction pipe. Usually the amount of water delivered

STEAM TRACTION ENGINEERING

to the boiler can be exactly regulated by partially or wholly closing a valve fitted to the by-pass pipe between the suction and discharge pipes. When starting the pump, first close this valve tight, then open it sufficiently to allow the required amount of water to be delivered to the boiler; the excess will flow through the by-pass back to the tank.

This pump is subject to the same troubles as the one referred to above and therefore will require the same remedies.

If the gearing on the main shaft and the small crank shaft are kept well oiled or greased, the pump will make very little noise and will give satisfactory service.

Clark Pump.—This is an independent steam pump which gets its power from a small vertical steam engine. (Fig. 16.) The pumping parts are almost identical with the two types already mentioned. The piston of the engine is connected by means of a large yoke to the pump plunger. The crank shaft is located within the loop, the connecting rod being very short. A small flywheel on the crank shaft carries the piston past the centers so that an eccentric can drive the slide valve in the cylinder. This flywheel usually has a small crank attached so that the pump can be run by hand. In case the water is too low in the boiler to start a fire safely, the pump can be operated by means of this crank without the necessity of removing a priming or filler plug of a boiler.

The water ends of these pumps are the same as those of the crosshead pumps and the same troubles and remedies will apply to them. The steam end will require the same attention as the main engine. The pump will give little or no trouble if the cylinder, valves and other working parts are well supplied with the proper kind of oil and are properly adjusted. The steam piston rod, also the valve stem should be kept packed with a good quality of steam packing.

BOILERS AND BOILER ACCESSORIES

The water plunger, on the other hand, will require water packing of good material.

Should the steam valve for any reason need to be reset, follow instructions given for an ordinary slide valve on page 194 with the one exception that in this case the lead wants to be twice as much on top as on the bottom.

If the engine is fitted with a feed water heater, the exhaust from the pump engine should be piped to it. Considerable annoyance and trouble will be saved if a check valve is placed between the heater and the pump.

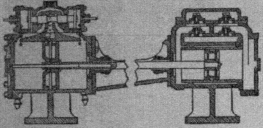

FIG. 17.—MARSH PUMP.

These pumps can be run just fast enough to supply the boiler under ordinary running conditions or may be speeded up, if so desired, to fill the boiler quickly, so that no controlling valve is needed on the suction piping or delivery.

Marsh Pump.—This pump is of the independent steam-driven type with no outside valve gear (Fig. 17). The water end, which is of the piston pattern, contains two suction and two discharge valves. Directly over the suction valves (the two lower ones) is the so-called water-valve plate which carries the discharge valves and is fitted on by means of a gasket. The suction valves divided into two compartments by a central wall are connected by means of passage ways above them to each end of the water cylinder. A cap with the air chamber attached is located over the discharge valves and is fitted to the cylinder by means of a gasket. The piston consists of a follower or head having a recess for packing and a follower plate secured by means of a nut to the small piston rod. The end where the piston rod works through has a packing gland and nut; the other end of the water cylinder is covered by a substantial head. The water enters the pump through the annular space around the water cylinder whose only connection to the cylinder is through the suction valves.

A description of the working of the pump in one complete revolution is as follows: Suppose the water piston is at the right-hand end of the cylinder; then as it moves toward the left it will leave a vacant space (partial vacuum) behind it. This space has no connection with the outside except through either the suction or discharge valves. The discharge valves will be held tight to their seats by their spiral springs but the atmospheric pressure on the water in the tank will cause it to rise in the suction pipe and force the suction valves off their seats, thus allowing it to enter the cylinder behind the piston. When the piston starts to return the right suction valve closes at once, due to the lack of a vacuum and to the greater pressure in the cylinder than in the tank. This same pressure will force the discharge valve off its seat, causing the water to flow into the air chamber and discharge pipe to the boiler. The same sequence of operations happens in the left side of the piston while it is making one revolution.

The steam cylinder is like that found in any ordinary steam engine with the exception that it is a little more than twice as long as the stroke of the pump. The steam piston, which is on the same piston rod as the water one, has a double head each of which is fitted with self-expanding

BOILERS AND BOILER ACCESSORIES

rings. The heads have a space between them called the trip space, which is used to hold the steam supply used for tripping the steam valve. As will be seen from the cut, the steam chest is fastened directly on top of the cylinder and is bored considerably larger at each end than in the middle. The steam valve, which is of brass, has large heads at each end used to trip the valve by means of live steam currents. The spaces at each end next to the heads are turned smaller than the valve body and constitute the means of admitting steam to the ports, while the smaller portion in the center controls the exhaust. The exhaust port is the center opening below the valve. As shown by the dotted lines, a small hole is drilled from each end of the steam chest into the cylinder; also one from the steam cavity in center of the steam chest through the hollow piston rod to the trip space. Holes are also drilled at the inner ends of the large chambers of the steam chest for the admission of steam to the steam ports. These small holes with the space between the heads of the piston are required in order that the valve may be reversed at the end of each stroke without the use of any mechanical connection to the piston or its rod.

The construction of the steam end of the pump having been given, the operation of this end of the pump can now be easily followed. By referring to the cut, it will be noticed that the piston is in the center of its travel and its valve is at its extreme right. Therefore steam can pass from the chamber above the valve through the small circular opening at the left end of the same into the cylinder, causing it to be pushed to the right. Steam also passes up through the small drilled hole at the left end into the valve chamber, thereby holding the valve in this position by equalizing the pressure on both sides of the valve head. When the piston reaches the end of its travel to the right, steam coming through the small drilled hole at this end to the right valve chamber overbalances the pressure on the

STEAM TRACTION ENGINEERING

other end of the valve by reason of the fact that the left valve head is in perfect balance. The valve will now be driven to the left, opening the right steam passages and the left exhaust port. This will cause the piston to move to the left and the same process will be repeated. It must be remembered that at all times the space between the two heads of the piston is filled with live steam supplied by the centrally drilled holes from the steam space.

On the side of the pump will be found a small plate containing the exhaust deflecting valve which is operated by a small lever from the outside. This valve is used to turn the exhaust steam into the water going to the pump for the purpose of heating it.

Small starting pins are placed at each end of the steam chest in order that the valve may be pushed by hand should it stop on dead center or refuse to start.

Before starting a pump see to it that the lubricator is filled with plenty of good fresh cylinder oil and that it is in good operating condition. Never allow the pump to run without oil, as this kind of treatment will soon cut the valve and cylinder.

In starting a pump first throw the exhaust lever towards the steam end of the pump before turning on the steam. After turning it on, open the small pet cocks under the steam cylinder until the pump has made a few strokes. It is a good plan to run the pump without water until it works easily and smoothly. When ready to begin pumping, open both the stop valves on the suction and discharge pipes. Next open the relief cock on the water end of the pump until the latter is working smoothly; then it should be closed. Now throw the exhaust lever back towards the water end of the pump, which will cause the exhaust to enter the feed water tank.

As a pump cannot lift hot water on account of the vapor filling the vacuum as fast as it is formed by the piston, the

BOILERS AND BOILER ACCESSORIES

exhaust should be turned into the tank only when the water is cold. Remember also that cold water is not good for the boiler.

These pumps can be regulated to supply the required amount of water which the boiler needs, by controlling its speed by means of the valve on the steam line to the pump.

Now that the construction and operation of this type of pump has been given, it is well to know the possible troubles that may arise and how they may be found and repaired.

1. If the pump runs but the water fails to flow from the tank to the cylinder of the pump, the trouble may be any of the following:

a. Suction hose and connections may leak air due to a break in them.

b. Hose may have collapsed or flattened by the suction.

c. Lining of hose may be loose or broken so that it interferes with the passage of the water through it. This is a common trouble with all pumps using a suction hose.

d. Screen on hose may be filled with dirt, thus choking off the water supply.

e. There may be an obstruction under the water valves that will not allow them to seat properly.

f. There may be leaks due to defective packing under the water-valve plate. If upon examination it is found defective a new gasket should be made and put in its place. This gasket must be patterned from the faced top of the water cylinder and the gasket between the valve plate and the air chamber must be patterned from the air chamber and not the valve plate.

g. Packing in the follower head in the water cylinder may be faulty or worn out. Special packing is required for this, which may be procured for a few cents already cut to size from any supply house.

STEAM TRACTION ENGINEERING

h. The follower may be loose.

i. Head of water cylinder may be loose or the gasket broken. Do not use heavy packing for this purpose; thin cardboard soaked in oil or thin sheet rubber will do better.

j. Packing may be worn out in gland around piston rod. Use candle wicking or asbestos for this and for the steam cylinder gland on rod.

2. If the pump takes water and then slows down or hardly moves, look for obstructions in the delivery pipe.

a. Check valve may be partially stuck or does not raise enough.

b. Pipes may be choked with lime. Examine the nipple where the water enters the boiler. It may be choked with lime so that the opening is no larger than a lead pencil. This is quite a common and frequent cause of trouble.

3. If the pump does not supply the quantity of water that it should at ordinary speed, it is due probably either to a leak in the suction pipe, a loose lining in the suction hose, water valves leaking or plunger packing worn. First examine the suction pipe or hose; if no leak or trouble can be found or if they have been adjusted, then examine the piston packing.

a. In repacking the piston screw the follower head up firmly. Care should be taken not to pack it tight. If properly packed the piston should be moved readily by hand.

b. If the water valves leak, they should be reground to a new seat. For instructions see page 91.

4. If the piston does not move when the steam is turned on, any of the following troubles may be the cause:

a. Steam does not reach the pump due to the inlet steam pipe being choked with scale and rust. In this case the pipe should be dismounted, examined and replaced after being blown clean with steam.

BOILERS AND BOILER ACCESSORIES

b. The steam valve may be stuck. It should be able to be moved easily. A little kerosene will loosen any rust or gum that may be present.

c. The head of the valve is loose.

d. The starter pins may be stuck or packed too tight. They should be able to be moved easily.

e. Water piston packed too tight. It should be examined and, if found necessary, repacked.

f. The small holes in the steam chest may be plugged up on account of the packing having squeezed out over them. This packing, which can be made of thin cardboard or heavy manilla paper, should be cut and patterned after the top of the steam cylinder and not that of the bottom of the steam chest. Be sure to duplicate carefully all holes in the packing exactly the same as the top of the steam cylinder.

g. The steam cylinder and valve have not been supplied with adequate quantity of good cylinder oil.

5. If the piston does not reverse at the end of its stroke, it is due to the following reasons:

a. The water piston follower head has become loose and worked itself either partly or completely off. This would prevent the piston from making a complete stroke and also would stop the pump by not allowing the valve to receive its trip supply of steam.

b. The small trip holes in the steam chest and the corresponding holes in the cylinder may be stopped up due to the gasket between the chest and cylinder having shifted. These holes are very important ones in the operation of this pump and it should be seen that they are kept entirely free.

c. The starter pins may be stuck by tight packing, thus not allowing a full valve movement. Pull them out as far as they will come and repack them at the first opportunity.

STEAM TRACTION ENGINEERING

d. The head may have worked off the steam valve. When replacing it screw it up snugly but be careful not to injure the valve or seat.

e. A thick ring of packing under the steam cylinder head will allow the piston to overrun its trip supply. Nothing heavier than manilla paper should be used for this gasket.

Injector.—This is an apparatus for forcing the feed water into the boiler which works on a principle entirely

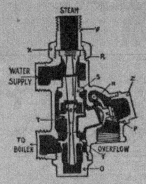

FIG. 18.—AUTOMATIC INJECTOR.

different from that of the pump. The fundamental principle of all injectors is that the velocity of the steam carries the air inside the injector with it and thus creates a partial vacuum into which the feed water flows. The steam then imparts a portion of its velocity to this water, giving it sufficient momentum to force open the check valves and enter the boiler.

While there are many different makes of injectors on the market they all work on the above principle, differing

BOILERS AND BOILER ACCESSORIES

only in minor details of construction. Of all the different makes the automatic injector shown in Fig. 18 is one very widely used on farm engines.

The operation of the above injector is as follows. Steam enters through the passage marked V and then through the steam jet R which has a very small hole directly over the suction jet S. The latter jet has a considerably larger opening. The steam rushing through these two jets creates a partial vacuum in the suction chamber W and suction pipe, permitting atmospheric pressure to lift the water into the chamber. The inrushing steam strikes this water and is partially condensed but it also imparts its velocity to the water, thus causing it to be driven forward with very great velocity into the boiler by way of the delivery tube Y, chamber O and the discharge pipe. The steam at first only drives out the air and as this cannot cause a high enough overpressure through velocity to overcome the boiler pressure, the air and steam are discharged through the overflow valve P. Through the medium of the small holes in the delivery jet a vacuum is created in the chamber T by closing the washer valve to its seat at the lower end of the suction jet. The force of the water causes it to rush past these small drilled holes at such a speed that the vacuum thus caused closes the overflow valve to its seat. The injector will not work properly or at all should the least bit of air enter it either through the suction pipe and connections or through a leaky overflow valve.

An injector is a simple machine having only one moving part (overflow valve) but the numerous small passages are required to be kept clean at all times.

Automatic injectors, such as the Penberthy, will automatically reëstablish the flow of water to the boiler should it be interrupted for a short time without the necessity of manipulating any valves. This is one of the reasons why this type of injector is used on portable and traction

STEAM TRACTION ENGINEERING

engines where it is liable to have many interruptions when going over rough ground.

An injector will not work well when the water is taken from a heater; besides little can be gained through its use as the water is heated in the injector nearly as hot as could be done with a heater. The latter will also cause considerable back pressure due to the large amount of piping.

The injector should have a separate feed pipe and check valve between it and the boiler.

When choosing an injector the following conditions should be considered and a selection made accordingly:

1. Distance water has to be lifted if the source of supply is below the injector.

2. The lowest and highest steam pressure carried in the boiler.

3. The temperature of the water supply.

4. The size of the boiler or the maximum amount of water required. The maximum capacity that the injector can handle should be thirty per cent more than the amount of water required. It is better economy to have a larger injector than is actually needed as hotter water may be delivered to the boiler by throttling the water supply.

A good injector should be able to accomplish the following:

1. Start with 20 to 25 pounds of steam with a 3-foot lift or less, the water supply being at a temperature of 74°.

2. Work with steam as high as 165 to 170 pounds pressure, the lift and temperature of water being the same as above.

3. Lift water as high as 20 to 24 feet with 60 to 80 pounds of steam, the water being 74°.

4. Handle water having a temperature of 125° to 130° with 60 to 80 pounds of steam, the lift being 3 feet or under.

The capacity of an injector is usually based on a 3-foot lift, a steam pressure of 60 to 110 pounds and a water supply temperature of 74°. The above capacity will be decreased with either very high or low steam pressures according to the following scale. With 30 pounds pressure the decrease amounts to one-quarter of the above, with 125 pounds it amounts to one-tenth and with 140 pounds it is one-quarter again. Increasing the lift will decrease the capacity in about the same ratio as that due to the increase in steam pressure. This capacity will also be decreased as the temperature of the feed water is increased, according to the accompanying scale. Water at 90° to 100° decreases the capacity one-quarter, at 115° to 120° decreases about one-fifth. Should the water supply be elevated above the injector, hotter water could be handled and there would be a slight increase in its capacity. Should the feed water temperature be less than 74° the capacity would be increased.

As dry steam will make an injector work much better it should be taken through a separate pipe if possible from the highest point of the boiler. A globe or angle valve should be placed on the suction pipe so that it closes against the water, i.e., the water should enter the valve from the under side of the disk.

The directions for starting a Penberthy, Lee, Garfield, Ohio, U. S., Lunkenheimer or Chicago automatic injector, which are all of the same general design and are operated in the same way, are as follows:

With less than 65 pounds steam pressure in the boiler, open first the water valve one turn; then open the steam valve wide. The injector now ought to operate perfectly but if water should run out of the overflow valve, it will be necessary to close slowly the water-supply valve until no water appears through the overflow. It sometimes happens that a mixture of hot steam and water will be discharged through the overflow valve, in which case the water supply valve should be slowly opened until the right amount of

STEAM TRACTION ENGINEERING

steam and water is present in the injector to make it operate perfectly without any waste through the overflow. It is only with experience with each injector and installation that the proper amount of opening of the water-valve to start the injector can be learned.

With over 65 pounds of steam pressure, open the water-supply valve, but remember that an injector will deliver hotter water to the boiler if this valve is allowed to be partly closed. Never regulate an injector with the steam valve but leave it always wide open and manipulate only the water supply.

In starting a Metropolitan automatic injector, the small handle at the end of the injector should be opened about half a turn, otherwise it is operated in the same manner as the other injectors. It may be stopped by means of this handle but as this causes it to get too hot, it is better practice to use the steam valve.

The failure of an injector to work may be for one of the following reasons:

1. A leak either in the suction pipe, hose or around stem of water supply valve. Forty per cent of all troubles are due to one of these causes. The suction line must be kept air-tight. To test for leaks plug the overflow and end of suction pipe or hose, then turn on steam, keeping a sharp lookout for the appearance of any leaks. If any are found they should be fixed at once.

2. Water supply cut off due to strainer at end of hose being clogged with sediment or a loose lining in the same.

3. A leaky check valve. The injector will become too hot to work and therefore should be cooled off by pouring cold water over it. The valve should be reground or replaced with a new one at the earliest opportunity.

4. Look for dirt in chamber where steam and suction jets meet, also in delivery tube. A small pebble or stick may work itself into the small holes in the delivery jet. All

BOILERS AND BOILER ACCESSORIES

these passages must be kept clean and free from dirt. Remember that no injector is expected to work with chaff and dirt nor take water from a tank that is full of trash and mud.

5. The water may be too hot for the injector or the steam pressure too low. The only remedy for this is to get colder feed water or an injector made especially for high or low temperatures.

6. The jets may become coated with lime, in which case they should be soaked over night in a solution consisting of one part muriatic acid to ten parts of water or, instead, use strong vinegar. It is not a bad plan occasionally to soak the whole injector in the above solution.

7. The steam valve may leak enough to keep the injector too hot to work. In that case the valve should be reground or replaced with a new one.

In case the injector lifts the water but does not force it into the boiler, the trouble is probably the obstruction in the delivery pipe or a small leak in the suction pipe.

a. The check valve may be stuck. A light tap on the under side of the valve may loosen it. Do not strike it a hard blow as that only damages it so that it becomes useless. If the light tapping does no good, it should be removed and an examination made. It is sometimes found to have a projection or burr caused by wear which can be removed by following the instructions given on page 89.

b. The check valve may not have a sufficient lift to allow the water to pass through quickly and easily. If this is the case get a new check.

c. The feed pipe may be clogged with lime where it enters the boiler. If it is, clean it out thoroughly.

d. The water may be forced through an old heater having coils which are partially clogged up with lime. This should be avoided if possible. To determine the condition of the feed line, temporarily place a steam gauge in the delivery pipe close to the injector. If this gauge indicates several pounds more than the boiler pressure before it "breaks," it is almost certain that there is some obstruction in the feed pipe to the boiler.

c. Dirt in chambers, spill holes of delivery nozzle or in strainer will act in the same way as a throttled water supply.

Sometimes an injector will "drizzle" at the overflow with pressures at which it used to work well. The cause of this is that the nozzles have become either worn or clogged. A leaky overflow valve may also cause this condition. First see if the trouble cannot be remedied by regrinding the overflow valve. If the trouble is not obviated then purchase and install new nozzles. Steam nozzles are subjected to very little wear but the delivery nozzles will wear very fast and require renewing often. A new set of nozzles will usually make an old injector as good as new.

Feed Pipes and Their Fittings.—All boiler feed pipes should be as short as can be made with as few bends as possible and should allow the water to enter the boiler as far from the fire surface and below the water line as can be done practically. The feed pipe should enter a locomotive boiler on the side of the shell about halfway between the fire box and the front tube sheet. In the case of return flue boilers the place of entrance should be on the side of the shell a little forward of the center and for vertical ones the bottom of the water leg is a good safe place to enter the feed pipe.

By using tees instead of elbows on the nipples where the feed pipe enters the boiler, the pipe can be cleaned of lime and other deposits by the removal of the plug without taking the pipe down. Brass plugs should be used at these places as they are not nearly as apt to corrode as the iron ones. A good substantial stop valve should be placed close to this tee and a check valve equipped with a drain cock

BOILERS AND BOILER ACCESSORIES

in the bottom of it must be placed as near as possible to the stop valve. Of these two valves the check is the one placed nearest the pump or injector. By this arrangement when the boiler is under steam the check valve can be examined or the pipe drained in cold weather.

If a heater is used a check valve having a bottom drain cock should be placed between it and the pump or injector and as close to the latter as can be done conveniently. Do not use a smaller feed pipe than the delivery connections on the pump or injector as this may cause trouble, especially with an injector, due to the considerable back pressure and friction. All feed pipes should be supplied with drain cocks at their lowest point to provide a means of draining the pipes in cold weather.

Steam pipes for injectors or steam pumps should be the full size of the connections. In order that dry steam should be obtained at all times for the pump or injector, it is advisable to obtain the steam from the dome of the boiler instead of from the water level. The steam pipe to an injector should not be a branch of another steam line supply as this sometimes makes the injector work badly when the other pipe is being used. An exception may be made to this in the case where the supply for an independent steam pump and injector can be taken from the same source when both are not apt to be used at the same time. Should this be done, have both branch pipes fitted with good stop valves, also one close to where the pipe is tapped into the boiler. This last valve is fitted for the purpose of shutting the steam out of the pipes in freezing weather; also in case anything goes wrong with the other valves. While not obligatory, it is a good policy to have drain cocks on these steam lines, as was suggested in the case of the feed pipes.

When both pump and injector discharge into the boiler through the same feed pipe a good check valve must be placed on each pipe before they are joined together, otherwise either boiler feeder would send back pressure through the one not working.

The check valves should be of the extra heavy pattern and only malleable iron or steel pipe fittings should be used. In purchasing washout plugs get brass ones with large square heads.

Valves.—It is not expected that light-weight standard valve or pipe fitting built for pressures of one hundred pounds or less, will withstand the strain of the high pressures carried on the traction and portable engines of today, as they will soon cause more trouble and loss of time than the difference in price between them and good high-pressure ones would warrant. Therefore in buying a standard article purchase only those that are guaranteed to withstand the maximum pressure which they will be called upon to carry instead of one that is structurally weaker and cheaper in price.

If a valve should leak a little, never try to close it hard or twist it tight. This treatment will shortly ruin the best made valve. There is probably a little dirt or scale between the disk and seat and by trying to force it shut the seat will be sprung or the disk will be roughened. By giving the valve a quick twist open and shut the obstruction will be blown out and it can be shut without any strain. The valve should not be closed tight when cold as the heat will expand it, causing it to jam and spring and begin leaking. If a valve is reground before it leaks very badly, it will be able to last very much longer and the job of regrinding will be shorter. The removable disk valves are very good and the disks can be easily replaced when worn.

CHAPTER II

CARE AND MANAGEMENT OF BOILERS

The care and management of boilers is one of the most important subjects of steam engineering. The success of an engineer depends almost entirely upon his knowledge of the management of both the boiler and engine. Carelessness and neglect of this branch of engineering is the cause of at least nine out of ten of all the boiler explosions and therefore the good reputation of an engineer depends on avoiding this in attending the boiler.

It is well to memorize the following statements, as upon them depends the proper working of a boiler:

1. Be sure that there is plenty of water in the boiler at all times. *Never guess* at the amount but *know exactly* how much is present. Two inches of water in the glass gauge is sufficient for running and will insure a much drier supply of steam than can be secured with a higher water level. Some engines work better with only an inch and a half. *Always stop* the engine with at least *an inch* of water in the gauge as it will sink quite a bit upon closing the throttle. This is caused by rising steam in the upper inches of the body of water moving to the steam pipe when the engine is running. When leaving the boiler for the night see to it that the gauge is filled two-thirds full of water in order to guard against loss of level by cooling, evaporation and leakage. If left for a longer period extra water will be required for the same reasons. If too much water is found upon returning to the engine the excess can be run off through the blow-off valve.

STEAM TRACTION ENGINEERING

2. Keep as nearly a uniform steam pressure as possible and avoid extremely high pressures, thereby avoiding the possibility of a rupture or explosion.

3. Avoid sudden changes in temperature.

4. All accessories should be kept in good shape and the boiler clean both inside and outside.

5. Before leaving the boiler for the night close the dampers and add enough fuel to keep the fire going until your return. The fire can be held for some time if a couple of shovelfuls of fresh coal are dumped just inside the fire door or several large sticks of wood are placed on the sides of the fire box. If the dampers do not fit tight, the smoke box door had better be opened a few inches. Under no circumstances open the ash pit doors.

6. By legitimate means try to prevent the steam from blowing off through the safety valve while running, as this tends to raise the water and may start it foaming or priming, which is always objectionable and should be avoided if possible. If it "pops" while the engine is at rest, first start the injector or pump to create a circulation, then cover or bank the fire with fresh fuel to absorb the heat and allow the steam free egress through the safety valve.

7. Get the best water obtainable and see to it that the supply tank is kept clean.

Starting a New Boiler.—After examining all the fittings to see that they are screwed up snugly, that all the plugs and hand holes are in their proper place, the grates in position and nothing in the flues, smoke box or stack, then the boiler is ready to receive the water. To do this remove the filler plug and start the pump. In order to make the filling easier open one of the gauge cocks or the blower valve in order that the air in the boiler will have an opportunity to escape. The boiler should be filled until there are about one and a half inches of water in the gauge, after which the pump may be stopped until the steam is gen-

OF BOILERS

proper amount of
this kind of fire gives
up evenly and allows a
out this time drops of
but they will disappear
warmed up thoroughly;
point. When the steam
advisable to tighten up on
this may save trouble later
gaskets blowing out when
pressure. The pump and
see that they are in proper
ween fifty and sixty pounds
they should not be in good
adjusted at once.
it has been used before, make
am pressure intended to be
re is any doubt on this ques-
tested before proceeding any
page 99.) In the meanwhile
if necessary clean them of scale
em with new gaskets. They
before the boiler was laid up
A thorough examination should
and connections to see that they
cial attention being given to the
gauge cocks and nipples where
boiler. These connections should
ot clogged up in any way with
re is any doubt as to the steam
ely it should be tested (page 95).
ve been attended to, the boiler is
I the directions given above should

in regard to the water level will not.

be out of place as upon this depend in a large measure the continuous and efficient operation of the engine and the safe condition of the boiler.

Keep as near as possible a constant supply of water in the boiler and never allow your vigilance to relax for a single moment in this matter. The water should be kept within a range of not more than one inch variation while running.

When the engine is running use the pump, setting it so that it will supply just the right quantity to the boiler. The injector is not capable of close regulation by throttling and therefore will be used only when the engine is stopped for any purpose. Should the boiler be provided with two injectors and no pump, use both for short intervals, rather than one only for long periods.

Firing with Coal.—In firing with coal keep covering the fresh and incandescent coal on the grates with thin layers of fresh coal thrown in at short intervals. Large pieces of coal should be broken up before being used. This is the most economical and satisfactory method of firing. By careful study and practice a good fireman can soon learn to control the fire and keep an even pressure of steam. Do not open the fire door and then go and get your coal—a bad practice which is, however, often followed—as this allows cold air to rush into the fire box against the hot sheets and tubes, causing seams, rivets and tubes eventually to leak. The door should be open only during the actual time of firing; the continual practice of keeping it open for longer periods may eventually ruin the best boiler ever made. In firing with some grades of coal, clinkers are formed on the grates which exclude the air, thereby reducing the draught. They should be removed, but not, if it can be helped, while the engine is running. The time for removing them and cleaning the fire is in the morning before getting up steam for the day's work, at noon, or before

CARE AND MANAGEMENT OF BOILERS

banking the fire for the night. Should it be necessary to clean the fire while the engine is running, do it quickly, shutting the door as soon as it is finished. As a fire may be started by the hot clinkers and cinders, it is well to keep a pail of water or a hose handy to extinguish it. For the same reason wet down a place before dumping the ashes.

The exhaust of the engine should provide ample draught while it is running except in the case where it is running light or the coal is of poor quality; then the draught may be insufficient to keep up steam. In this case the blower may be used to create the required amount. In getting up steam it is inadvisable to use the blower with less than twenty pounds in the boiler but after this pressure is reached it is of the greatest value in raising steam. With seventy-five pounds or over the blower should be only partially opened in order to give the best results.

When leaving the boiler at noon, fill it as for the night, only there is then no need of banking the fire with as much coal. The poker should not be used any more than is absolutely necessary while the engine is running as stirring the fire will cause clinkers to form.

When burning coal in a straw-burning boiler one or two fire bricks are of benefit by causing better combustion and a steadier and more even heat.

The safety valve should be allowed to blow off at least once a day to see if it agrees with the steam gauge, but frequent opening of the valve is an indication that the engineer is careless or cannot control his fire.

Firing with Wood.—With wood instead of having a thin fire, the fire box should be kept full all the time by the addition of a few sticks at short intervals. The wood should be placed as loosely as possible so as to allow the flame to pass freely between the sticks. The sticks can be placed in alternate rows across and lengthwise of the fire box. In burning "slabs" a better steaming fire will be obtained if the sticks are placed sawed side down instead of the bark side. Large round sticks will burn much better if they are split before being placed in the furnace.

A wood fire will require an occasional knocking down with the poker, but do not try to stir all the hot coals through the grates as they are a great source of heat and will give a much better fire when allowed to remain.

To produce the same results, wood does not require as much care to get the required air as coal. For this reason it is an advantage to use a dead plate when burning the former. This is a solid plate from six to ten inches wide extending the full width of the fire box and is placed on the grates at the furnace door end of the fire box.

A screen or a good spark arrester should be placed in the stack as an incandescent spark of wood is lighter and therefore more apt to be carried out of the smokestack than in the case of coal. When threshing around buildings, near heavy stubble or in fact any place where there is danger of fire, notify the owner or employer of the possible danger present, thereby putting the responsibility of damage on his shoulders. While any burning wood is apt to give sparks, yet it is especially true when compelled to use rotten wood or old rails. Whenever the screen becomes burned out it should be replaced with a new one as soon as possible and not neglected.

Firing with Straw.—The modern straw-burning boilers are of the direct flue type except in a few cases where a return flue is used. The fire door is removed and a straw chute is attached in its place. In direct-flue boilers a fire brick arch is used which extends from just below the flues against the tube sheet, upward and backward to within ten or twelve inches of the back end of the fire box. This arch will become white-hot and will therefore materially help the combustion of the fuel. The draught is taken in

CARE AND MANAGEMENT OF BOILERS

at the front end of the fire box, as the grates extend only about half its length, the remaining space next to the fire door being occupied with a dead plate.

When firing with straw keep the chute full all the time by the addition of small forkfuls of the fuel, but do not push it into the fire box faster than it can be consumed, as the fire may become choked and give very poor results. By keeping the chute full, cold air will be prevented from entering the boiler by this way. The flame as seen through the peephole should appear white-hot where it comes over the arch.

A medium fine mesh screen must be placed over the smokestack to prevent the live cinders from escaping into the outside air. The screen may be choked up after the engine has been standing idle for some time, thus interfering with the draught unless the cinders and soot are removed. Allow all but sixteen inches in the front end of the ash pit to be filled with ashes. With this space ample draught will be provided and better results obtained than with a clean ash pit. When burning wet or rotten straw the draught is often insufficient for good combustion, in which case the blower can be used to considerable advantage. The exhaust nozzle may have to be reduced in order to give sufficient velocity of draught, but this should never be done unless absolutely necessary, as by doing this a back pressure will be produced in the engine.

Sometimes either chaff, short pieces, damp or rotten straw will be drawn over against the tube sheet and must be removed by scraping off with the poker thrust through the peephole. Occasionally turn the fork over and run it under the straw in the chute into the fire to break up any bunches that may be formed on the grates.

In order that the engine should steam well the boiler should be level or a little raised at the end farthest from the fire. The tubes should be cleaned at least once a day

STEAM TRACTION ENGINEERING

and the front or stack damper should always be the one used in connection with the one under the fire door.

Never be discouraged if more fuel is required one day than the preceding day, as atmospheric conditions, such as the change of the wind, the amount of dampness in the air and the change in the temperature, will largely influence the amount of fuel required. Again, some waters will require more fuel than others to produce the same amount of steam. A great deal depends on whether the boiler is clean or dirty, full of mud and scale or recently washed out.

A good fireman will always be trying to better himself by seeing if he cannot run on the succeeding day with a little less fuel, while a poor or indifferent one will be continually poking his fire, cleaning the grates or putting in fuel all the time, thus making hard work for himself and not producing any more steam for the large amount of fuel that he is burning.

Grates.—Burnt grates are caused by carelessness and neglect on the part of the fireman in allowing the ash pit to remain filled with hot cinders or coals instead of attending to their removal. If engineers would only give the grates half the attention that they deserve there would very seldom be a case of burnt grates. They are only caused by such fuels as coal or wood. Burning straw will not form a bed of hot coals on the grates as well as under them; therefore no trouble should be experienced with the grates from this cause. As has already been stated, it is advisable to allow the ash pit to remain filled with the hot ashes from the burnt straw. When the ash pit is allowed to be filled with hot incandescent coal or wood fuel the cool air is prevented from passing through the grates and the air which does enter becomes heated and carries acid from the ashes so that the hot coals on top of the bars cause warping or burning by reason of the excessive heat. As

CARE AND MANAGEMENT OF BOILERS

long as the air can circulate freely under the grates and up through their openings, a hot enough fire cannot be built to ruin them. Large hot clinkers which are left lying on the grates a long time will cause the top of them to become burnt or rough, allowing the clinkers to stick fast and become hard to remove.

Rocking grates should be kept in good shape or else they will become worse than the ordinary ones. They should be freely movable at all times. When at rest they should be kept level; if tilted up on one side the high projections will burn off, thereby spoiling them. Long grate bars will sag much quicker than short ones; for this reason grates which are divided into two sections with a supporting bridge in the center are not apt to become sagged even if overheated, but they are open to the objection of burning off at the center support unless great care is taken to keep them free from clinkers at this point.

Examine the grates frequently and remove the clinkers which have attached themselves either to the bars or in the spaces between them. Coal sometimes melts and runs down between the bars, where it is next to impossible to remove it except by entering the fire box and doing it with a chisel and hammer.

When putting in new grates do not allow them to be crowded tightly between the fire box sides, as the expansion will cause considerable strain on the fire box side sheets. They should be at least three-quarters of an inch shorter than the length of the fire box, otherwise the grates will warp and sag because they cannot expand without binding. The size of the grate openings can be larger when burning wood or lump coal than is possible with good results when using a finer size of fuel.

Tubes.—Unless the tubes of a boiler are kept free of soot it will be a hard matter to keep up the steam pressure. There are a number of different makes of good flue cleaners on the market. If the engine is not already provided with one, as it should be, it is advisable to purchase one as soon as possible. The cleaner should be used every morning before starting the fire, thoroughly to clean each of the tubes. To get the best results the cleaner should fit the inside of the tubes snugly.

Do not choke the exhaust nozzle and try to make the engine clean its own tubes, as it will cause considerable danger from fire and also reduce the power of the engine.

Straw and some other kinds of fuel will cause a kind of gummy soot to adhere to the surface of the tubes which is hard to remove.

Cleaning a Boiler.—No specific interval of time between boiler cleanings can be given, as it all depends upon the kind of water used, the amount of work the engine is doing and the thoroughness with which the preceding cleaning was done. When using hard water that forms scale easily the boiler should be cleaned as often as every three or four days, while with soft water it may not require cleaning for two or three weeks. In general, using ordinary water, this work should be done once a week. When a boiler begins to foam it is a sure sign that it is in need of a cleaning. Do not try to run with a dirty boiler, but clean it often enough to prevent the mud and sediment from filling up the water spaces around the sides of the fire box.

In cleaning a boiler which has been running all day, do not attempt to blow it out, but wash it with a good hose and force pump. When the engine is at the end of the day's work, take every bit of fire out of the fire box and clean the ash pit of all cinders and ashes; next close the fire and ash-pit doors, and allow the boiler to cool down. If the boiler is to be used again, blow off the water with about ten pounds' pressure, but do not fill the boiler again or try to wash it while hot, as the quick and uneven contraction caused by the cold water striking the hot boiler sheets is largely the

CARE AND MANAGEMENT OF BOILERS

cause of so many leaky boilers. The blowing off process should be used only when necessary, as it will cause the mud to become baked to the inside of the boiler by its contact with the heat of the iron. The baked mud is hard to remove, and this method should be avoided if possible. A better plan is to prepare the boiler for cleaning on Saturday and allow the water to remain in it overnight. Wash it the next day after the water has been drained with a hose, as the mud and sediment will then be soft.

In washing out a boiler the following method should be followed:

After removing all the hand holes, attach a hose having a good nozzle to the tank pump. Wash the crown sheet, the upper part of the boiler and under the flues. For this purpose there are located a hand hole in the back head above the crown sheet and another in the front tube sheet under the tubes. If the boiler is of the water-bottom type there are from two to four hand holes located at the lower corners of the same, while if it is of the open-bottom type their location is at the lower corners of the water leg. These should be removed so that the water can have free egress from the boiler. The hose should be continued to be used on the boiler until the water leaving it through these hand holes is nearly clean. The feed pipe, water gauge and gauge cocks should be examined for the removal of possible deposits. After washing out the boiler the hand holes should be replaced and the boiler refilled with fresh water for future service.

In replacing the hand holes it is advisable to replace the gaskets with new ones. An exception can be made to the former statement in the case where the gaskets used are of metal. Care should be taken to replace the metal gasket in exactly its former position. Clean the inside edge of the hand holes and plates of all old packing or scale that may adhere so that the new gasket will have a smooth surface to press against. A good gasket can be made from an old rubber two-ply belt, or it can be cut from sheet rubber, which is procurable from any of the well-known supply houses. One of the best materials to make hand-hole gaskets is a red rubber tube, which can be cut to the proper length and the two ends joined together by means of a small lead tube which comes with the packing. The gasket ends should be cut on a diagonal so that they will lap and this joint wrapped with a short piece of electrician's tape. Oil should be placed on the gaskets so that they will not adhere to the boiler; also the bolts and nuts should be oiled so that the plates can be tightened up easily. The plates should be tightened up securely when the boiler is cold and again tightened when there is twenty to thirty pounds on it. This little attention will prevent the gaskets blowing out at a very inopportune time after the boiler becomes hot.

It is sometimes very hard for the gaskets to hold on an old boiler, in which case the following built-up form of gasket, made by alternating layers of thin sheet rubber and ordinary window-screen gauze, may be tried. First, a screen gasket is placed on the hand-hole plate, on top of which is placed a rubber one, then another screen, the latter of which has its surface coated with either white or red lead, and over this is lastly placed another rubber gasket. As stated before, the hand-hole plate and outer face of gasket is oiled to prevent sticking. If this form of gasket does not hold, the only remedy will be to use a metal one.

Metal gaskets should be cut out of lead sheets, about one-eighth of an inch thick, free from holes or blisters. They can be used over and over again, provided care is taken to replace them in exactly the same position that they were formerly every time the hand-hole plates are removed.

Foaming and Priming.—A boiler may foam and not

CARE AND MANAGEMENT OF BOILERS

prime, but it is rare for it to prime without first foaming. A boiler will foam when there are present certain kinds of floating foreign substances or scum which appear on the surface of the water. These substances are the soluble minerals present in the water which are liberated in the process of boiling. The liberated steam bubbles must force themselves through this scum which, being lighter than the water, lies upon its surface. Water which is alkaline or contains gypsum is very bad in this respect. Stagnant water taken from ponds full of moss or slime is to be avoided if possible on this account. As stated above, a dirty boiler is liable to foam when out on the road because mud is stirred up within it. The presence in the water of greasy, oily or soapy substances is very bad. Never allow the water tank to be cleaned with soap, as even a small quantity of this will make the boiler foam.

When a boiler is foaming the water will appear oily or milky in the gauge glass and the water level will change rapidly, the glass appearing nearly full one minute and the next instant almost empty. It is, therefore, very important when this does occur that the proper amount of water in the boiler be accurately determined. In order to do this it may be necessary to close the throttle for a few minutes to let the water settle.

The best remedy for a foaming boiler is to wash it out as soon as possible. If it is fitted with a surface blow valve, blowing off at this point for a minute or so may help to decrease the foaming by removing some of the scum from the surface of the water.

Priming is the lifting and carrying of large quantities of water through the steam pipe into the cylinder of the engine. This will cause the engine to appear to be working very hard, exhausting heavily and throwing a great deal of water out of the smokestack. The presence of water in the cylinder can be heard by a loud clicking noise. The cylinder head may be knocked out or the engine otherwise seriously damaged by the water in the cylinder. The water washes all the oil out of the cylinder and off the valve seat, causing the valves and cylinder to work hard.

Priming may be the result of any one of the following causes:

1. Foaming.
2. Too much water in the boiler.
3. Engine working hard with front end of boiler low.
4. Valves improperly set. (For instruction on resetting see page 194.)
5. Leaky valve or piston.
6. Boiler working beyond its capacity.

Should priming occur, close the throttle partly, enough to slow down the engine, and open the cylinder cocks until dry steam again appears, when the speed of the engine may be slowly increased until it has attained its normal speed. When priming is caused by an overload of the engine, the only remedy is to reduce the load or get a larger engine. When a boiler is priming a melted fusible plug or a burnt crown sheet is liable to result unless a close watch is kept on the water level.

Scale.—It is a well-known fact that all waters used for boiler purposes, unless they are rain or distilled water, contain more or less scale-forming materials in solution. The materials which form incrustations on the inside surfaces of boilers are for the most part carbonates and sulphates of lime and magnesia in the feed water; they are precipitated when the water is heated.

Scale in the boiler is one of the indirect causes of boiler explosions (page 99). The most dangerous feature of scale is that the conditions brought about by it are not manifest until after the damage is done. Usually the evidences of scale in the running of a boiler are but indications that damage has already been done. For instance, a tube be-

CARE AND MANAGEMENT OF BOILERS

gins to leak as a result of overheat caused by an accumulation of scale on its outside or water surface. Overheating a tube injures it; therefore the leak proves that the injury has already been done. The tube may be reëxpanded and made tight for a while, but it will be the first one that will have to be renewed. Again the first indication that a crown sheet is overheated is the presence of leaky stay bolts or a bagged sheet, but the crown sheet was injured at the time it was being overheated. In each of the above cases the damage done is the first indication that dangerous conditions were in existence. A man who does not realize the danger of boiler scale is apt to wait for some sign of weakness or trouble, and such a man takes his life in his hands and the lives of those who are working near the engine. He is not a safe man to be in charge of a boiler, no matter how careful he is with machinery. Many engineers unfortunately do not realize the danger and loss that is caused by boiler scale. If they did, the scale would be taken out of the boilers as soon as possible. Good engineers sometimes operate boilers containing considerable scale, as when taking charge of a boiler that has been neglected, but they get rid of it at the first opportunity and see to it that it does not again accumulate. In order to operate a boiler safely and with the best economy, it is necessary to know the following three facts concerning scale:

1. Where will it accumulate in the boiler?
2. How fast will it accumulate?
3. What is necessary to get rid of it?

No one can tell without making a close examination whether there is a large or small amount of scale in a boiler.

Scale will form in the most out-of-the-way places in the boiler, and no accurate estimate as to the condition can be made from the way it looks inside of the hand holes. Among the places that are liable to fill up with scale are the pipes

STEAM TRACTION ENGINEERING

connecting the water column, if one is used, water glass and gauge cocks. The same trouble will occur where the water glass and gauge cocks are attached directly to the boiler. Starting with a perfectly clean sheet, pipes and openings it will take some time for them to become choked, but after a thin coating is once formed the process will be more rapid because the roughened surface catches and holds the small particles of scale-making material carried in the water. The rougher the surface becomes, the more rapid will be the process of incrustation in and around the openings. When the glass or column is blown out or the gauge cocks used, the water, rushing outward, carries small particles with it, part of which adhere to the sheet and pipe at the opening, eventually closing it. These fittings must be removed in order to clean the openings and connections of deposits. A boiler which is fitted with a water column should use tees with plugs instead of elbows on its connections, so that a rod can be run through the pipes upon the removal of the plugs and the opening cleared in this manner. Such a cleaning is not as thorough as it should be, however, because the scale is not removed from the sheet, thus leaving a very rough surface for further accumulation. Do not think because the full area of the pipe is restored by means of a rod that it will take as long as it did before again to be choked with scale, for such is not the case, and by presuming so, dangerous conditions may be present before the engineer is aware of them. An opening cleared in this manner still retains its rough and uneven surface, upon which new particles of scale will easily attach themselves. Fig. 19 illustrates such a pipe opening. A shows the condition before the rod is used; B condition immediately after using the rod. It is to be noted that the rod has not touched or loosened the scale which has adhered to the sheet on both sides of the opening. C shows how the new particles of scale have again obstructed the opening.

CARE AND MANAGEMENT OF BOILERS

The rapidity with which scale will accumulate in one part of a boiler cannot be judged by looking at the amount present in another part of it. This is one of the peculiar features of boiler scale and is a reason why all parts of a

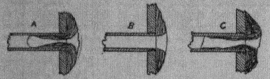

FIG. 19.—DIAGRAMMATIC VIEW OF BOILER SCALE IN TUBES.

boiler should be examined thoroughly each time it is cleaned. Make a complete examination every two months and keep track of the thickness of the scale adhering to any of the surfaces and openings in the boiler during this

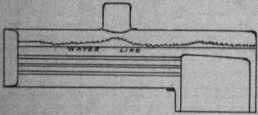

FIG. 20.—DIAGRAMMATIC VIEW OF SCUM ON SURFACE OF WATER.

interval. Knowledge of this fact will be a good guide in determining how often inspections should be made in the future.

When a boiler is under steam a more or less thick scum or froth will cover the surface of the water, as shown in Fig. 20. This scum is largely composed of scale-making ma-

STEAM TRACTION ENGINEERING

terial, which will get into the steam openings where it is baked on their surfaces by the steam. This requires that these openings be looked after as well as those of the water. While this scale is not as hard as that formed below the water surface, a piece of it will become brittle and tough enough to clog effectually the upper gauge-glass opening or steam-gauge siphon. Should this happen serious trouble may result from a wrong reading of these important gauges.

The accumulation of scale in the crown sheet and around the staybolts, if allowed to remain, will become heavy and

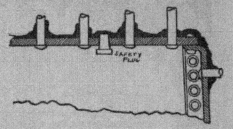

FIG. 21.—DIAGRAMMATIC VIEW OF SCALE ACCUMULATION ON CROWN SHEET AND STAYBOLTS.

solid. (Fig. 21.) The water will therefore be kept away from the sheet and the former will not be allowed to absorb its heat nor that of the staybolts' heads. If allowed to persist in this condition, the sheet and staybolts will become overheated. An overheated staybolt will become brittle or break, leaving the boiler in a very dangerous condition. They are almost sure to leak when the scale is removed, the boiler being in the above condition. The same thing will occur with the rivets in the fire box seams. The overheating of the sheet covered with scale may also cause the sheet to form small cracks around the staybolts and rivets. (Fig. 22.) The places of thickest deposit are where the water currents are abruptly changed. Therefore scale will accumulate on the tube ends, as illustrated in Fig. 23.

Notice the distance between the water and the thin metal

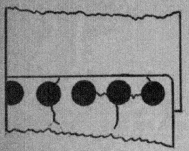

FIG. 22.—DIAGRAMMATIC VIEW OF CRACKS AROUND RIVETS AND STAYBOLTS.

of the tube where it is expanded in the tube sheet. As a result the tube will become overheated or warped, making it leak when the scale is removed, or sometimes even before.

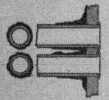

FIG. 23.—DIAGRAMMATIC VIEW OF SCALE ON TUBE ENDS.

Engineers often blame the kerosene or boiler compound for these leaks, but it can be readily seen that the scale is the direct cause of them.

STEAM TRACTION ENGINEERING

The constant contraction and expansion of a boiler often causes the scale to flake off in either large or small pieces. These will accumulate around the fire box sides, especially in open-bottom boilers, often to such an extent that they will fill up above the fire line. The lower part of the fire box sides will become weakened through being overheated.

No rule can be laid down as to how often inspections, cleanings or blowing down should be done, because such periods depend entirely upon conditions. Good judgment and carefulness are the only rigid rules to apply in the handling of boiler scale. The danger due to scale is not unavoidable. It does not require much time or very hard work to find out if scale is present or is likely to be, but it does require a little dirty work. It is better to have soiled overalls than a ruined boiler or an explosion.

Where scale has accumulated it may be removed by mechanical means. These are the hand or power tube cleaner, scalers, chisels, hooks or chains. Many of these tools can be made by the engineer.

Boiler scale can be prevented from forming by the addition of certain compounds to the feed water. There are three methods which are used to prevent scale from forming a hard adhesive crust or coating.

The first of these is to introduce in the boiler some reagent or material which shall prevent the scale from hardening or crystallizing by a sort of mechanical reaction.

The second method is to make use of some material in the boiler which will act as a varnish caked upon the surface of the boiler so that scale will be less certain to adhere.

The introduction of kerosene or starch operates in this way. The former is one of the best scale preventives that can be suggested, provided the refined product is used and it is handled carefully. A quart of this material

CARE AND MANAGEMENT OF BOILERS

should be poured into the boiler after it has been cleaned and before starting to fill the boiler with water. As the water gradually fills the boiler, the oil will adhere to the scale and plates. It will cut loose the old scale and prevent the new scale from sticking to the sheets and tubes. A boiler which is badly scaled should receive a pint of kerosene every day, introduced with the feed water. This can be easily done by attaching a tee to the suction pipe of the

FIG. 24.—COMPOUND FEEDER.

pump or injector and using the device shown in Fig. 24. The oil is poured into the chamber when the pump is well started. The valve should not be allowed to remain open when there is no oil in the chamber as the air will enter the pump and cause it to stop working. If the boiler is in good condition a pint every other day will be sufficient, but should the water be very bad a pint every day is not too much. If kerosene is used very little scale will appear at the end of a season's run, provided the boiler is cleaned once a week and the best water obtainable used for the boiler feed. Crude oil should never be used in a boiler as it will form a very tenacious scale of its own. Some authorities consider that crude oil is injurious to a boiler, but this statement cannot be said to be true for all of them.

STEAM TRACTION ENGINEERING

The third method is to introduce a reagent which shall act chemically on the precipitate either to change its crystallizing or solidifying character or to change an insoluble into a soluble salt. Common soda added to water containing sulphate of lime will prevent the formation of hard scale. The water so treated will have the sulphates changed to carbonates, causing a soft scale which is easily removed by a good washing out of the boiler. If an excess of soda is used, the boiler is apt to foam. One-quarter to one-half a pound per day is usually sufficient but sometimes a quarter is considered an excess. Among the acids used for scale prevention may be mentioned acetic and tannic as the most usual ones, while carbonate of soda, chloride of barium, tannate of soda or sodium triphosphate are the salts that are used. The amounts of these to be added to the water can only be determined on the basis of chemical analysis. For the reason that farm engines usually obtain their feed water from a number of different sources during the course of a season's run it is particularly hard to tell which of the salts should be used. If the engineer is getting all his water from one place during the season, he should have the water analyzed and a particular compound prepared expressly for its use.

Low Water.—Allowing the water to get low in a steam boiler not only injures it but leaves it in a dangerous condition. In nine-tenths of all cases where it occurs, it is due to carelessness and inattention on the part of the engineer. As stated before, a melted fusible plug or a burnt crown sheet or both will result from allowing the water level to become low. The crown sheet will be burnt should the fusible plug become coated with lime. In this case the plug will not melt until the water level has dropped enough to expose the uncooled crown sheet to the hot gases. A burnt crown sheet is a very dangerous element in a steam boiler.

CARE AND MANAGEMENT OF BOILERS

To be on the safe side, stop the engine when there is no more than an inch of water in the gauge glass and no more water coming. Some engineers will keep on running the engine until the boiler is nearly dry, when they are expecting a water wagon to arrive any minute. This should never be done as the wagon may break down when only a short distance away from the engine. It is better to lose fifteen to twenty minutes with the engine shut down, waiting for the wagon, than to lose half a day putting in a new fusible plug. It will take a week to fix a burnt or bagged crown sheet. By refusing to run with low water, the engineer shows good commonsense as well as a realization of the danger. Low water is one of the most common of the indirect causes of boiler explosions. (See page 64.)

If the water in the gauge glass is discovered to have suddenly disappeared, the engine being idle, do not start it nor the pump, as the commotion thus created will throw the water over the hot sheets and is apt to cause an explosion. Also do not try to draw the fire, for the latter will be stirred up in trying to remove it, thus creating an intense heat. This is exactly what the engineer should avoid. Instead of doing these things, cover the fire as quickly as possible with ashes or dirt and leave the boiler alone until it has had time to cool down. Do not touch the engine, pump, nor any of the steam outlets, as a serious accident may be caused. When the boiler has cooled down a close examination should be made of it and any leaks that may have developed should be repaired. If there is any reason to believe that the crown sheet has become very hot, it would be advisable to have the boiler tested before further use.

If it is discovered that there is no water in the gauge glass or bottom gauge cock when the engine is running, it is necessary at once to get the front of the engine

STEAM TRACTION ENGINEERING

higher than the rear. To do this run the rear wheels into a ditch or the front ones up on a coal- or wood-pile. If the engine is on soft ground, holes may be dug under the rear wheels. This work must be done quickly. If after this has been done there is water in the gauge glass or cock, the pump or injector may be started and allowed to run until there is a normal supply in the boiler. By getting the front of the boiler high, the water will be allowed to cover the crown sheet unless it is exceptionally low.

If no water appears in gauge glass or lower cock after the front of the boiler has been raised, then it will be necessary to cover the fire heavily at once and allow the boiler to cool down before trying to do anything else with it. The fusible plug should be examined to see why it did not melt and be replaced with a new one. The water may be below the gauges and still have covered the crown sheet, but there is no way to tell this absolutely except by guessing. The word "guessing" should not be in the vocabulary of an engineer working around steam boilers.

Boiler Explosions.—The direct cause of boiler explosions is a pressure of steam in it greater than it is designed to withstand. There are two ways in which this may occur:

1. The boiler may be too weak for the pressure, so that it ruptures at its working pressure.

2. The pressure may become too strong for the boiler to withstand and it ruptures at some point above the working pressure.

The boiler may be weakened through improper construction, improper installation and incompetent or careless operation. While improper construction is not the usual cause of ruptures in boilers today, nevertheless it may be present, especially in old boilers. A boiler having

CARE AND MANAGEMENT OF BOILERS

lap joint seams is a dangerous one compared to one having these seams butt-jointed. The lap joint is dangerous because this form of construction promotes the formation of incipient cracks in the upper surface of the lower lap where they may be impossible to detect. Poor workmanship and abuse of material come under this heading.

The improper attaching of the usual boiler appurtenances such as the safety valves, steam and water gauges, check, blow-off and stop valves are indirect causes of boiler rupture.

Nine-tenths of all boiler explosions come under the heading of incompetent or careless operation. A list of these causes would be a repetition of those things which the authors have been trying to impress upon the engineer in the first part of the book. But in order that there should be no mistake as to what they are, they will be repeated.

1. Allowing the steam gauge to get out of order.
2. Allowing the water-gauge connections to become so clogged as to indicate ample water when there is none in the boiler.
3. Allowing the safety valve to become so glued to its seat as to fail to blow at the pressure for which it was set.
4. Allowing grease to enter or scale to accumulate in the boiler, on the heating surfaces.
5. Allowing large quantities of cold water to impinge against the hot plates.
6. Allowing water to be driven from the heated surfaces by forced firing.
7. Allowing a large valve to be opened too suddenly.
8. Allowing minor repairs to be neglected until they endanger the whole structure.

A boiler will rupture or explode due to the pressure being above the normal working point and too great for

STEAM TRACTION ENGINEERING

it to withstand, due to either one or both of the following causes:

1. Steam gauge out of order or not giving a proper reading.

2. Seat of safety valve glued fast or valve set to blow off at too high a pressure.

Many violent boiler explosions occur either just prior to starting the engine in the morning or while they are idle at the noon hour or shortly after they have been shut down for the day. The reason for this is that heat will accumulate rapidly in the boiler if steam is not being drawn and should the safety valve fail to relieve the excess pressure an explosion will soon follow.

Preparing a Boiler for Storage.—When through with the use of the engine for a considerable period, as from one season to another or for two or three months, precautions should be taken to prevent deterioration by rust or oxidation of the plates both inside or out. Instead of using a convenient fence corner for storing the engine, it should be protected from the weather by a good shed.

The outside should be cleaned of all grease and dirt, after which all exposed sheets, including smoke box and stack, should be painted with a good coat of boiler paint to prevent rusting. If the boiler is jacketed the jacket should be rubbed with a piece of rag or waste soaked in machine oil. The tubes, fire box, smoke box and stack should be cleaned of all deposits and soot. The ashes and cinders should be removed from fire box and ash pit. It is a good plan to remove the grates and give them a thorough cleaning. The boiler should be washed thoroughly to remove all of the deposits possible. Then it should be filled full with water, leaving only room enough to pour two to two and a half gallons of kerosene on top of the water. In order to do this the whistle or safety valve may be removed. Now draw the water off slowly

CARE AND MANAGEMENT OF BOILERS

through the blow-off valve. When it has all been drained off remove the plug, should there be one in the bottom of the boiler, and also the hand-hole plates. These should be left off in order that the air may circulate freely through the boiler, thereby keeping it in the best possible condition until it is again to be used. Drain all pipes, pumps and gauges. If the engine is not to be used for a considerable period, the brass trimmings ought to be removed, cleaned and packed in a box. The old packing should be removed and thrown away. A good deal of trouble caused by old packing will thus be avoided in the future.

It will only take a day or two to do this work and will lengthen the life of the boiler, besides earning the satisfaction of having it in good shape for next season's run. A boiler which is not given any attention may give out after three or four years when it ought to last twenty if it is rightly treated. Also by doing these things when the run is over, anything that needs repairing, such as leaky tubes, staybolts, worn fittings or valves, pumps, injectors or piping, is more readily noticed and should be put in order at this time. If not done now some of the weak places will be forgotten when it is time to start the boiler at the beginning of the next season. These weak places will eventually cause trouble, necessitating laying up the boiler for repairs right in the busy part of the season when the engine should be running every day.

CHAPTER III

THE REPAIRING OF BOILERS

Tools Required.—Before an engineer can do properly any boiler repairing work, he should provide himself with a set of appropriate tools. These are a good ball pein machinist's hammer of about one and a half pounds' weight, a tube expander of the proper size, a hack saw,

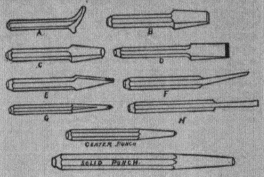

FIG. 25.—BOILER REPAIR TOOLS.

beading tool, two calking chisels, several types of cold chisels and a punch. Other tools may be added to the equipment as time and circumstances require. The above tools can be either purchased or made by the engineer.

The beading tool should be shaped like A in Fig. 25,

THE REPAIRING OF BOILERS

being dressed out, slightly rounding in the heel or gullet, so that there will be no sharp edge to cut the tube. Do not make the J toe or guide too slanting or the tool will work badly. This tool should not be tempered too hard. In tempering, heat the tool to a very bright red, just beginning to appear white, and then plunge it into a can or receptacle containing ordinary black or machine oil. It should not be withdrawn until cold.

A square joint calking tool, shown as B, is used for slight leaks and a finishing run over a joint after the round-nose tool C has been employed. These tools are best when made out of three-quarter-inch tool steel about six inches long. They should have as nearly as possible the slant shown in the cut. The round-nose tool will not work well if it is too thick and the edges on the end merely rounded off. The object of having the edges rounded is to avoid cutting or burring the sheets. When tempering calking tools, cold chisels and punches, heat them to a bright cherry-red and plunge them into heavy oil. They should be kept in the latter until cool. With a little practice an engineer will soon be able to temper the tools to any degree of hardness that he desires.

It is desirable that the tool kit should contain at least four different kinds of cold chisels. For ordinary work the chisel shown as D can be employed. A handy one to have for a good many places is E. This is commonly called the "cape chisel." The one illustrated under F could be called a "flue chisel." It is used to advantage for cutting off old tubes inside of the tube sheet. The "diamond point" chisel G will be found very handy in removing a piece of broken off pipe from valves and fittings. A ripping chisel or "ripper" is shown at H. In the line of punches only the center and solid ones are illustrated.

In addition to these tools, the following extra ones will

be needed for such work as applying a patch and putting in staybolts: a staybolt tap, a patchbolt tap, a boilermaker's ratchet and several different size drills. On account of the cost of a staybolt tap, it need not be purchased but instead can be borrowed usually from almost any machine shop.

Putting on New Tubes or Flues.—Farm engines are more subject to tube trouble than all other boiler weaknesses put together, and therefore every engineer should know how to re-tube one. Bad water will often cause them to burn off at the fire box end. In fact the water in some localities may be so bad in this respect that the tubes will not last more than two seasons.

The proper tool to use when cutting the ends of an old tube preliminary to removing it is what is known as a tube-cutter. This instrument is costly and is not apt to be in the possession of most engineers. If borrowed, it should be adjusted so that it will cut the tube off *inside* of the tube sheet. If this is not done the tube sheet will be ruined and also perhaps the cutter. Do not try to make the cutter do the work all at once as either the cutting wheels may break or the feed-screw spring. Take all the time necessary to do the work right. In locomotive boilers cut the bottom tubes first so that they can be removed through the hand hole in the front tube sheet. Should there be no hand hole at this point it would be well to cut such a one (see page 87). In return tubular boilers the upper tubes should be cut first and removed through the hand hole over the main flue.

Without the tube-cutter the job will be somewhat longer and harder as it will be necessary to resort to the hammer and flue chisel. In order to prevent a nick or rough place being made in the tube hole, the tube should not be cut off too close to its sheet. Such a nick will cause trouble when putting in the new tubes. It is desirable

THE REPAIRING OF BOILERS

to use a sharp cold chisel or square-end calking tool in cutting off the bead in the fire box or combustion chamber. To do this place the tool against the outer edge of the bead, the point towards the center of the tube. This will slightly curl the end of the tube inwards, thus loosening it and at the same time cutting off the bead. Next cut the tube off just inside of the smoke box tube sheet, this being the end opposite that from which the bead has been removed. Push this end over to the hand hole and pull the tube out. Should the tube stick fast in the other tube sheet, place a calking tool against the end which is fast and give it a few light blows with the hammer. When doing this do not allow the sheet around the tube hole to be cut or bruised by the tool. If copper rings or ferrules are used around the tubes at the fire box end, they should be removed and later replaced with new ones. Proceed in the same manner with each tube until all have been removed from the boiler. To remove the short pieces left in the tube sheet, place a square-end calking tool under the bead and turn it up and inwards. The piece then can be bent towards the center of the hole and easily driven out.

Both sides of the tube sheet should be cleaned free of all soot and scale after the removal of the tubes. The same should be done with the tube holes. If by accident any holes in either sheet should be roughened, they should be carefully dressed smooth with a rat-tail file. Take care not to increase the circumference of the hole under any circumstances.

Little details like these may seem insignificant but attention to them means thorough work and therefore should never be neglected.

If the new tubes are too long, they will have to be cut before being installed. If the engineer has no regular tube-cutter a hack saw should be used. For beading allow about three-sixteenth-inch projection at each tube sheet. This

STEAM TRACTION ENGINEERING

amount will not make a large bead but one which will be easier to turn and is not so apt to split. A large one will burn off much quicker than a small one, because of the increased amount of metal in contact with the fire. It is always a good idea to have at least one extra tube to be used in case one should prove defective or should be spoiled in the course of installing it. If copper thimbles are used in the fire box tube sheet, do not try to expand the tubes in the holes without replacing them with new ones. The tube will be split unless they are used. If these thimbles are not available, get a sheet of copper and cut it in strips just wide enough to lap over the tube sheet slightly, i.e., nine-sixteenths inch wide for a one-half-inch tube sheet. It is well to provide some sheet copper, even if thimbles have been obtained, in order to bush some of the holes which may have been stretched by former expanding. Always bush a tube in a hole where a piece of copper can be inserted that will encircle half of the tube. The copper used for the above purpose should be about the thickness of twenty-gauge iron. If one strip is not enough use as many as is found desirable.

Before starting to put in the new tubes provide an iron rod a little longer than their length—a tube scraper handle will do—a can of oil, some rags or waste, expander, light ball pein hammer, beading tool, copper thimbles, sheet copper and a hack saw. Place the tube in the holes by sliding it over the iron rod. After allowing the tube to extend three-sixteenths of an inch beyond the tube sheet, mark it plainly so that no mistake will be made as to the proper place to cut off the tube. Now remove the tube and saw it off at the mark as nearly square as possible.

To avoid getting a tube in the wrong hole at one end, always start at a corner and run the rows straight. Be careful about this as tubes that are put in wrong will have to be removed. Now place the tube back in position and

THE REPAIRING OF BOILERS

slip the thimble, if one is used, in place. If they are not used and the tube is a little loose in the hole, apply as much "bushing" as can be slipped in easily.

In expanding a tube probably a better job can be done by using a roller expander. This tool is also easier to handle. Insert the expander in position inside the tube with the guide or shoulder up snug against the tube sheet, taking care to keep the tube in its right position with three-sixteenths of an inch of it sticking through the tube sheet. See to it that it does not slip while working upon it. With the hammer drive the tapering pin of the expander in the tube just a small distance. Next turn the pin around several times, using a small rod placed in one of the holes provided on it for this purpose. Repeat the driving and turning until you have the tube tight in the hole. The tube should not be rolled too hard or it will be cut off or become so thin that it will not hold for any length of time. The amount of rolling a tube needs is something which every engineer will have to learn by experience. It is better far to have to re-roll the tubes after testing them than to roll them so hard that they are ruined. Another trouble due to rolling too much is the danger of springing the tube sheet or stretching the hole. Always keep the expander well oiled when using it or the pin may break.

In beading down the ends of the tubes, light blows with the ball pein hammer should be struck working from the center of the tube outwards. Turn the projections all the way around gradually downward towards the tube sheet instead of making it flat in one place all at once. When it has been turned down well towards the sheet, place the guide of the beader inside the tube and gently hammer the bead evenly until it is driven down snugly against the sheet.

Proceed in the same manner with all the tubes in both tube sheets. When the work is completed the boiler should

STEAM TRACTION ENGINEERING

be tested to locate any leaking tubes. Should the test indicate an appreciable leak, the faulty tube should be rolled a slight amount more. Never do this, however, when there is any pressure in the boiler, as it may cause other tubes to leak while tightening the defective one. A very slight leak will usually stop as soon as the boiler has been fired on account of the expansion of the tube and sheet.

Replacing Staybolts.—This is a ticklish job requiring some skill with tools and a great deal of common sense and judgment. A new staybolt will be required should one be found broken or leaking to such an extent that it cannot be stopped in the usual way. In order to make a good job the new bolt should be slightly larger than the old one. This is done to allow new threads to be cut in the old holes which will just fit the new bolts. This increase in size of bolt is usually about three-sixteenths of an inch.

With a sharp cold chisel cut the head of the old bolt away so that the edges can be seen. Punch a distinct mark with a center punch in the middle of it. This mark should be deep enough so that the drill can be started without slipping. Next proceed to drill the bolt by means of the ratchet and a drill bit one-sixteenth inch smaller than the size of the bolt. In order to do this it will be necessary to arrange some kind of solid block for the ratchet to work against, with a small piece of iron fastened to the block for the feed screw to abut against. The plate must have a heavy-center punch mark in it to prevent the ratchet screw from slipping sidewise. While doing this plenty of oil should be used on both the ratchet and drill. The hole should be made about five-eighths of an inch deep. The bolt should be easily pried out with a punch after the above work has been done on both ends of it. Should it not be easily removed cut it off inside of the sheet from the inside of the hole by means of a diamond point chisel.

THE REPAIRING OF BOILERS

If any burrs remain in the holes they should be removed, care being taken not to cut into the sheet while doing it.

The engineer is now ready to tap the holes for the new bolt. In doing this be sure that the staybolt tap is kept well supplied with oil during the whole time that it is cutting. In order to prevent the tap from breaking it should be turned steadily and all jerking and sidewise pulling should be avoided. Always screw the tap through both sheets from the same side. If one hole is tapped out from one side and the other hole tapped from the other, then the threads will be apt not to coincide.

After tapping the holes the new bolts are ready to be put in. They should be at least one and a half inches longer than the distance through the holes from the outside of one sheet to the other, in order that there will be room enough to place a wrench on the end of the bolt and also to provide a projection through the sheets for riveting down. Insert the bolt and screw it through until it projects beyond the opposite sheet about three-sixteenths of an inch. With the hack saw cut off the bolt at the end on which the wrench was used, leaving the same amount of projection on each end. Rivet one end of the bolt using a ball pein hammer for the work, while another person holds a heavy sledge against the other end. Keep the head of the bolt as nearly round as possible. When well riveted down take a round-nose calking tool and go around the edges of the bolt. Follow this with light blows on the square-end tool to make a neat and tight job. Never try to rivet over a staybolt or do any heavy calking without a good heavy iron or sledge being held against the other end, as it is almost sure to loosen the bolt in the threads, causing it to leak.

In order to do this work well, never be in a hurry. The men who are able to command the highest wages and the choice of jobs are those who can be depended upon in a tight place to repair an engine in a bad condition, so that the repair parts are as good as new.

Tightening Leaky Tubes.—If the leak is slight, it may often be stopped by using only the beader, but should it be extensive, it is advisable to use both expander and beader. This is true provided the tubes are not burnt. Before any work is done the tube and tube sheet should be cleaned of all soot and foreign substances.

Oil the expander well and insert it in the tube, placing the shoulder up against the tube sheet. Drive the pin in lightly and turn the pin around several times. Repeat the operation until the tube is tight. No set rule can be laid down as to just how tight to expand a tube as this can be learned only from experience. When the tube has been rolled sufficiently, strike the pin a side blow to loosen it, then give it a few turns. Repeat this process until the pin can be easily removed. When all the tubes which have been leaking have been rolled, they should be gone over with the beader. Tap this tool, its guide being inside the tube, gently all the way around each of the leaking tubes. If the leaks fail to stop the first time try the above work again. If rolling and beading will not stop them or if they hold only a short time, it is safe to say that the tubes either have been burned or have been rolled so often that the metal of the tube is too thin to stand the strain and new tubes will be required.

Repairing Leaks at Staybolts.—If the bolt is not burned nor the sheet cracked the following method should be employed to stop staybolt leaks. If the leak is slight first go around the edge of the bolt with the round-nose calking tool and hammer, finishing the job with a square-nose tool. But if the leak is considerable, have someone hold a sledge or heavy iron bar on the opposite end of the bolt while driving up the head that leaks with a ball pein hammer. This will swell the bolt in the sheet and by finishing with

THE REPAIRING OF BOILERS

the calking tools as before stated, the leak will be closed tight. If in calking or riveting the bolt seems brittle and cannot be made to hold, the only remedy is to replace the old staybolt with a new one (see page 74).

When the sheet is cracked a different method must be followed (Fig. 26). If the crack extends from one bolt to another, the best plan is to cut out the defective part of the sheet and staybolts, apply a patch and then put in new

Fig. 26.—Cracks Around Staybolts.

bolts (see page 79). If the crack is a short one, it may be effectively stopped by the method known as "screwing up a crack" (Fig. 27). To do this take a seven-sixteenths-inch drill and ratchet and drill a hole in the crack so that one edge of the drill will cut into the staybolt slightly. Thread this hole with a half-inch tap and after threading a piece of soft iron or steel rod, screw it in the hole. Allow this to stick through the sheet about one-fourth or three-eighths of an inch on the inside and project from the sheet about one-eighth of an inch. Drill another hole in the same manner, allowing it to cut slightly into the inserted plug. Thread and insert this plug in the same manner as was

STEAM TRACTION ENGINEERING

done with the first one. Proceed in this manner until the crack has been entirely plugged. The last hole should have its outer edge extend slightly beyond the edge of the crack. When all the plugs have been screwed into place and cut off to the proper length, rivet them down with a ball pein hammer. Finish the job by using the calking tools on both staybolts and plugs.

FIG. 27.—PLUG INSERTED TO STOP STAYBOLT CRACK.

A cracked tube sheet may also be repaired in the same manner.

While this process takes time and skill, it makes a very satisfactory repair for small cracks and costs very little except the labor to do the work.

Leaking Seams.—Boiler seams will sometimes start to leak on account of the continual contraction and expansion of the sheets. When this occurs, the leak may be stopped by calking first with the round-nose tool and then finishing up with the square-end one. Where the leak is bad or extends for more than a couple of inches, it is a good plan to take a sharp chisel and dress about an eighth of an inch of the edge of the sheet away. The tool should be kept at a bevel

THE REPAIRING OF BOILERS

of about thirty degrees. This dressing of the sheet gets the edge away from any ridge formed by previous calking and presents a new clean surface to the work. When calking a leaky seam, always start an inch or so beyond the leak and with light blows work towards the center of the leak. Start on the other side and again work towards the center. This process is less likely to start other leaks than any other method which could be used and does not tend to extend the old leak any further.

Applying a Patch.—Never try to patch the crown sheet in a fire box boiler, because it will very seldom prove satisfactory in any way. The proper remedy for a burned or bagged crown sheet is to put in a new sheet and this requires the services of an experienced boiler-maker. A competent engineer should be able, however, to patch successfully the fire box sides, the main flues in return flue boilers and weakened shells.

In the case of traction and portable engines, the method of applying a patch by means of patch bolts, may be considered as the best, quickest and simplest method to employ. As an example of this method, there is illustrated (Fig. 28) the patching in the main flue of a return flue boiler.

Plainly mark on the old sheet the part which is to be removed, avoiding sharp corners, as they are very hard to calk tight. With a cape chisel cut a kerf out of the sheet along this mark. With this same tool next cut through the sheet at some convenient place and then use the ripper to cut out the defective piece. After the removal of the plate, true up the edges of the sheet, so that they will be as true and straight as possible. The patch should measure four inches more both in length and width than that of the piece removed. In Fig. 28 if A is line of the part removed and the dotted line B the patch, then the distance between A and B should be two inches. The patch should be located on the *water side* of the flue, because if it is put on the

STEAM TRACTION ENGINEERING

other side, it will leave a place for scale and sediment to lodge and burn it. The form of the patch should be such that it will fit snugly against the old sheet. It would be easier to handle and less liable to fracture while shaping it, if it is heated. The sheet around the hole should be thoroughly cleaned of all old scale, dirt or rust. The center of the bolt holes in the old sheet should be on an imaginary

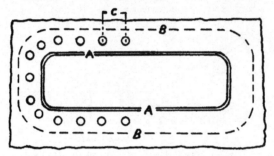

FIG. 28.—APPLYING A PATCH TO A MAIN FLUE OF A BOILER.

line drawn one inch from A. If using five-eighths-inch patch bolts, the distance between their centers—called the pitch and marked C in the cut—should be between one and one-half and one and five-eighths inches. Make a good punch mark in the center of the bolt positions and drill a half-inch hole in the sheet. Now place the patch in position and carefully mark on it the corner holes to correspond with the holes in the sheet. These corner holes should then be drilled, care being taken that they match the corresponding holes in the sheet. Bolt the patch in place with common bolts through each of the holes in it, and proceed to drill the other holes in the patch, using the ones in the sheet as guides. This will prevent the patch from crawling sidewise.

THE REPAIRING OF BOILERS

A bolt should be kept in the hole next to the one that is being drilled to prevent it from springing. After all the holes have been drilled, remove the patch in order to remove all burrs from the patch and sheet. The holes in the sheet must now be reamed out large enough to allow the patch bolts to pass through and they must be also countersunk to fit the tapering heads of the bolts. If any burrs are formed during the reaming of the sheet they should be removed at once. The patch can now be replaced and the

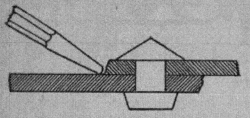

FIG. 29.—CORRECT METHOD OF HOLDING A CALKING TOOL.

holes in it threaded with a patch bolt tap. The tap should only be screwed in far enough to make enough threads so that the bolts will fit tightly. Insert all the bolts and screw them up evenly all the way around the patch. Never tighten one bolt as far as it will go and then do the same thing with the next, but keep going around the patch tightening each bolt a half a turn until all of them have been drawn up snugly. A small nick should be made in the neck of the bolt between it and the head where the wrench fits on and the bolts should be screwed until the heads are twisted off. In calking the edges of the patch, first take a sharp cold chisel and chip the edges of the sheet. Be careful that the patch is not cut into while doing this as it will make it hard to calk besides making a rough job. Next

STEAM TRACTION ENGINEERING

go around the joint with the round-nose calking tool using medium blows of the hammer on the tool. Finish this job with the square-end tool. In the same manner the bolt heads should be calked (Fig. 29).

This kind of a job if it has been done carefully should be entirely satisfactory in every way. Always test a boiler after the job has been done to see if any leaks appear.

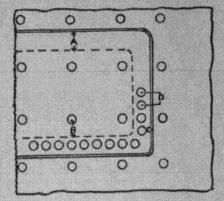

FIG. 30.—APPLYING A PATCH ON THE FIRE-BOX SIDE OF A BOILER.

If there are any the job will have to be recalked, using only the two calking tools. But do not do this, however, with any pressure on the boiler, as while tightening up on one leak another leak is apt to spring somewhere else.

The job of applying a patch to a fire-box side will be found to be a little more difficult than the preceding one, on account of the staybolts, which have to be encountered. As before mark off the piece to be removed as shown by the dotted line in Fig. 30. The distance B should be about one-half inch. With a cape chisel and ripper cut out the defective part along the marked line in the same manner as was described in the preceding job of patching. A patch of this kind will be much handier if placed on the fire side, allowing the patch bolts to be screwed into the old sheet. When the defective part has been cut loose, drill out the staybolts at their opposite ends according to the method described on page 74. The ends need not be removed from the piece of defective sheet, as it would entail a lot of needless work. Upon the removal of the defective sheet and staybolts, the edges of the old sheet should be dressed up and trued. The patch should lap over the edge of the hole about two inches (A in the cut). The bolt holes should be laid off as follows: The pitch line should be one inch from the outer edge of the patch to the center line of bolts, as shown at C, and the pitch (D) should be from one and a half to one and five-eighths inches. With a half-inch drill drill all the holes in the patch. Place it in its proper position and carefully mark the corner holes on the sheet. Drill these holes in the sheet carefully with a half-inch drill and ratchet. Do not allow the drill to "crawl" from its proper position, so that these holes will match evenly with the holes in the patch. Bolt the patch in position and using the holes in the patch as guides, drill the remaining holes in the sheet. Next lay off and drill the required staybolt holes in the patch, using a drill one-eighth inch smaller than the size staybolt required. After this ream and countersink the patch bolt holes in the patch, being careful to ream them true and allowing just enough clearance to pass through a five-eighths-inch patch bolt. The patch is now ready to be put in its proper position and bolted. Thread the patch bolt holes in the sheet and put in the bolts in their proper place, tightening them in the same manner as was stated above. Since the patch is on the outside it is calked instead of the sheet as was done

STEAM TRACTION ENGINEERING

in the case of patching a flue. Before finally fastening the patch in place with the patch bolts, do not forget to clean the sheet and patch of all burrs caused by drilling and also of all old rust and scale. For putting in the staybolts, proceed exactly as described on page 74. After the job has been finished, the boiler should be tested and if any considerable leaks have developed, it should be calked again. If a patch is required on any other part of the boiler, proceed in exactly the same manner as for these two patches already mentioned, as nearly as the nature of the patch will permit.

Another method of applying a patch is by means of rivets. This method is by far the best, but on account of the necessity of having to use a heel bar to drive the rivets tight, it can only be applied with great difficulties to certain parts of the boiler. One of the places where this method can be used is the front end of the boiler shell, where the front trucks are bolted in place. This kind of a patch is easily applied at this point on account of the hand hole in the front tube sheet, through which the rivets can be inserted and held in place while driving them up. Proceed in the same manner as described above to cut the defective sheet and make a patch to fit it after allowing two inches of lap. The patch should be shaped to fit the old sheet snugly all around. Do not forget to straighten and true up the edges of the old sheet. It is well to use five-eighths rivets. The pitch should be one and five-eighths inch with the pitch line at least one inch from the edge of the patch. Lay off and drill the holes in the patch, using a twenty-one-thirty-second drill for a five-eighths-inch rivet. For this work always use a drill one-thirty-second of an inch larger than the rivets. Put the patch in position, mark the corner holes in the sheet and drill them accurately. Next bolt the patch back in place with five-eighths bolts and drill the remaining holes in the shell, using the

THE REPAIRING OF BOILERS

holes in the patch as guides. Remove the patch and clean it of all old scale and rust, also remove all burrs that might have been caused by drilling. Again replace it in position and apply the rivets. For this purpose there will

FIG. 31.—HEEL BAR.

be required a heel bar (Fig. 31) which is simply a heavy iron bar with a recess in one end to fit the rivet heads. This end must be bent so that it will go in at the hand hole

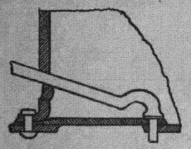

FIG. 32.—USE OF A HEEL BAR.

and at the same time set squarely on the rivets. Fig. 32 illustrates how this bar should be used. The rivets should be about three-fourths of an inch longer than the distance through the sheets to allow plenty of metal for riveting on a head. Before inserting a rivet in one of the holes it should be heated nearly white hot. It should be riveted over while still hot, being held in place by the heel bar. From the above it will be seen that a helper will be necessary to do this particular kind of patching. There must be someone to insert the rivet and hold the heel bar solidly on the rivet head while the engineer is riveting the other end as quickly as possible. Proceed in this manner until all the rivets have been driven into their places. The bolts in four corners of the patch should be replaced with rivets. In doing this work be sure that all the heads fit snugly against the sheet, before driving the rivet. Never drive a cold rivet; if it turns black before it gets into place or before it fits snugly, knock it out and use another one. Never try to use a rivet that has been bent or battered. After all the rivets have been driven, the edges of the patch and the rivets are ready to be calked. But before doing this dress the edges of the patch, using a sharp chisel at an angle of not more than thirty degrees. Care should be taken that the sheet is not cut while doing this dressing. In calking use the round-nose tool first and finish with the square-pointed one. Test the boiler as before and tighten any leaks that may appear when there is no considerable pressure on it. In applying any kind of a patch, always use boiler steel for it. Plow steel or iron should never be used, as it is very hard to work and is never satisfactory.

Rivet Leaks.—If the leak is not excessive it can be stopped by calking. Should the rivet be in such a position that a heel bar cannot be placed adequately over its head, or if such a bar cannot be obtained, then it will be necessary to do the work in the following manner, attempted only after it is found that the leak cannot be stopped by any other method: Cut the head off of the defective rivet and drive it out with a good solid punch. Take a patch bolt and tap one-eighth inch larger than the rivet hole and after care-

THE REPAIRING OF BOILERS

fully reaming a countersink to fit the taper of the patch bolt, thread the hole and insert the bolt. Make a slight nick in the neck of the bolt with a chisel and then screw the bolt until the head is twisted off. With the calking tools calk the head of the bolt. The job is then finished and if carefully done will be satisfactory.

If the defective rivet can be reached from the inside, so that a heel bar may be used, the old one may be replaced with a new one in the following manner:

The old rivet is removed by cutting off its head and driving it out with a punch. A new one of the proper size to fill the hole and three-fourths of an inch longer than the thickness of the sheets is heated to a white heat and thrust into the hole quickly. While riveting, it must be held securely in place by means of the heel bar, the latter being held by a helper. If the head of the rivet is not held firmly against the inside sheet, it may be found necessary to do the job over again.

A new rivet may have to be inserted in case the old one is loose in the hole and it cannot be tightened sufficiently by calking to stop the leak.

Cutting a Hand Hole.—It is not advisable to cut hand holes in boilers unless absolutely necessary, as the boilers will be weakened to some extent. However, sometimes it is a great convenience to have one below the tubes, especially in the front tube sheet. Boilers are usually not provided with any at this particular point. If one is desired here, proceed with the work as follows:

First lay off and mark the hand hole in its proper position. The size may vary, but for most places an oval two and a quarter by three and a quarter inches will be found about right. Be sure to make it a perfect oval. The hand hole should be cut at least one inch longer than the width. With a diamond point chisel, take a light chip out of the sheet along the mark all the way around the plate. This will prevent making a mistake in cutting due to the erasure of the mark. Continue cutting with the diamond point until the hole has been cut, taking care not to spring or stretch the sheet. After the removal of the piece of sheet, dress the edges of the hole with a cold chisel. For the hand hole plate, take a piece of heavy boiler plate and cut two pieces out of it, one just a trifle smaller than the hole and the other enough larger, so that it will extend over it about three-eighths of an inch all around. Place the smaller plate concentrically over the larger one and rivet them together with about three-quarter-inch rivets. The bolt, of one-half-inch diameter, should be placed in the center of the plate. It is fastened by threading the hole and bolt and screwing the latter in tight before riveting. The yoke may be made from a piece of iron three-eighths inch thick and one and a quarter or one and a half inches wide. In cutting its length allow enough material for the proper bend desired and drill or punch a hole in the center for the bolt to pass through.

In tapping an extra hole for a pipe in any part of a boiler, cut it with a diamond point chisel, somewhat smaller than the size of tap to be used. Before tapping the hole, ream it with the proper size reamer. Never tap a feed pipe in the fire box sides, in a corner or at an abrupt turn of the sheets.

Fitting Gauge Glasses.—It often happens that the glass tube of a water gauge will break while the boiler is under steam, in which case a new tube will have to be inserted. To do this close both valves quickly, first the lower and then the upper, remove the packing nuts and old packing, cleaning them thoroughly. Then place the packing nuts on the new tube, slip the new rubber gaskets on, shoving them into place in the nuts. After this has been accomplished tighten the nuts just enough to prevent leakage, open the upper valve slightly and allow steam to blow through for

THE REPAIRING OF BOILERS

a few seconds in order to warm it evenly. The last step in the process is to close the drain cock and open both valves slowly until they are wide open. Should the packing leak, close both valves before tightening the nuts. If the engineer should attempt to tighten the nuts while pressure is on, he may get either scalded or cut by flying glass.

It is advisable to keep in stock an extra supply both of gauge glasses cut to proper length and rubber gaskets of proper size for packing. To cut a glass tube to any desired length any of the following methods may be employed. With a glass tube cutter, insert the wheel in the tube to the desired point, exert a slight pressure on the cutter while turning it by hand completely around. After removing the cutter the glass will break at this point upon gently tapping it. If not provided with this tool, take a three-cornered file and break off the tip, insert the above in the tube and scratch a circular mark by means of one of the sharp corners. Gently tapping the glass at this point will cause it to break evenly. A third method is to use the soft end of a match soaked in water. By placing the end of the match inside the tube and revolving it, a film of water will mark the place at which the glass is to be broken. Now hold this spot over a fire for a few seconds and the glass will break of its own accord at the mark. The gauge glasses should be wrapped separately in paper or canvas and kept in a box. It is not good policy to string the glasses on a wire as the slightest scratch will often cause them to break. They can be toughened by boiling in salt water.

Gaskets can be cut from sheet rubber. Either asbestos or candle wicking may be used as a substitute.

Grinding Leaky Check Valves.—Many dollars and much valuable time can be saved by knowing how to regrind and refit a leaky check valve. They are very easy to repair and are usually the ones that first need repairing. The valves are generally beyond repair, should they have

89

become badly sprung either by freezing or twisting when put on.

To regrind a valve it should be removed from the pipe and the cap taken off. After this has been done the valve should be cleaned as far as possible of all dirt and scale. If the guide on cup pattern flat seat valve, which protrudes upward inside of the valve cap, is fitted with a slot so that a screw-driver bit can be applied, rotate the valve on its seat by means of a brace and bit. But if this stem or guide should be three or four-cornered with no slot, a piece of pipe about a quarter-inch in diameter and three or four inches long with notches cut in one end to fit the stem can be substituted in place of the screw-driver bit. Place the body of the valve in a secure position so that it will not move, put a coat of oil and fine emery dust on the valve and valve seat and turn it rapidly for a few minutes on its seat by means of the brace. Should there be no emery available, grindstone grit, fine sand or gritty soil mixed with water or oil can be substituted. Now remove the valve, clean it and the valve seat with a rag or piece of waste. If both valve and seat appear bright all the way around, they have probably been ground enough to make the valve tight. But if either part does not appear bright around their whole circumference, they will require more grinding with a fresh mixture of oil and emery. This process should be repeated until the valve is ground down all the way around to its seat. Before screwing the valve cap on, the valve should be removed and both it and its seat cleaned of all dirt and grit caused by the grinding. Test the valve by trying to blow through it. If this can be done it will need a little more grinding. Sometimes a valve will become so badly worn that no amount of regrinding will make it tight. In such a case the only remedy is to buy a new valve.

A little rough place on some part of the valve is the

THE REPAIRING OF BOILERS

cause of a valve refusing to close when the current is stopped, allowing it to back out of the boiler. A light tap on the under side of the valve with a wrench or hammer will usually remedy this trouble at once. Care must be taken not to hit it too hard or it will either be bruised enough to cause it to leak or ruined beyond repair. If this occurs often, at the first opportunity take off the cap on the valve and remove the check. Before doing this if there is any steam in the boiler, close the stop valve between the check and the boiler in order to prevent the person removing the check valve from being badly burnt. With a piece of fine emery cloth rub the upper and lower guide prongs until all the roughness has been removed and the valve no longer sticks. In doing this remove only a slight amount of metal and be careful not to cut the seat at all.

To regrind a swing check valve, remove the small angle plug and put a little mixture of oil and emery dust on both the seat and valve. By means of the brace turn the valve on its seat until the valve seats properly. Proceed in the same manner as with an ordinary check valve described above. A ball check valve cannot be reground in the field because special tools are required which are not available outside of the factory.

Never have the wrench on one end of the check valve while screwing a pipe into the other end, as this treatment will ruin the best valve ever made. This twists the valve, thus throwing the seat out of shape so badly that it cannot be reground or repaired.

Grinding Globe and Angle Valves.—Standard globe and angle valves are made in two distinct styles, one being made with the disk and stem rigid while the other has the disk fastened to the stem by a ball and socket joint. The latter style is much the best for valves of three-quarters of an inch or larger. Under this size the valves are usually of the rigid pattern.

STEAM TRACTION ENGINEERING

To regrind a valve of the rigid pattern, remove the stem and disk by unscrewing the valve cap. Upon removing the hand wheel, the stem can be unscrewed easily from the cap. Put some emery dust and oil between the disk and its valve seat and turn the former vigorously on the latter for a few minutes by means of a brace on the stem. Always be careful, while doing this, that the stem stands in a true vertical position corresponding to the one it occupies when the valve is together. After the stem and disk have been turned a few minutes, remove them in order to clean the disk and seat and see if they have been ground completely around to a proper new bearing. Repeat the grinding process if necessary, until a satisfactory bearing has been formed. Test the valve by putting it together and blowing through it. If this can be done it will require a little more grinding.

The loose disk valves are ground in exactly the same way, but the disk must first be fastened to the stem. To do this unscrew the little jamb nut on the disk, remove the latter and place a small chip of wood in the socket and replace the jam nut in its proper position. A bad job is almost sure to result if care is not taken when fastening the disk to see that it is at right angles with the stem. After grinding, remove the chip so that the disk will again be loose on the stem.

Some valves are fitted with a small slot in the bottom of the socket. These may be ground by rotating the disk. Regrinding is out of the question with the high-grade valves fitted with a renewable disk and seat. The disks on these valves are made of composition, while the seats are of hard brass, and therefore unless damaged by freezing or twisting the latter need seldom be removed. To put in a new disk in one of these valves, unscrew the small nut on the disk holder, remove the old one by lifting it out, put the new one in its place and screw up the nut. Very often

THE REPAIRING OF BOILERS

the valve can be made tight by simply turning the disk over until it makes a new seat. If the valve is not made tight with a new disk, then a new seat will probably be required. To remove a seat, unscrew it by placing file or similar tool against the lugs to be found on its inner circumference. A new seat should never be screwed in very tight.

When putting a valve of any kind on a pipe, always place the wrench on the former, on the end which is being screwed. This will prevent springing or twisting the valve, which would be the case if the wrench was placed on the opposite end. This rule should be invariably followed with all valves and fittings.

Grinding Stopcock.—If the cock is sprung or bruised until it leaks badly, it will be almost impossible to grind it tight and a new one should be purchased. But if the leak is slight, due to wear, it may be reground again until it is made tight.

Take the cock apart and clean it out thoroughly of all scale and dirt. Cover the plug with a thin coat of emery dust and oil, put it back in the cock and replace the washer and nut. Screw the nut up with a small wrench until the plug turns a little tight. If the nut will not screw up sufficiently to draw the plug down, take a file and cut the shoulder on the bottom of the plug down a little. Also a small washer may be needed between the brass one and the nut. In doing the work be careful that the threads are not twisted off. After drawing down the nut, turn the plug around and around by means of a wrench. Continue this process for several minutes, meanwhile keeping the plug tight by screwing down the nut on the bottom of it. This should be continued until a new bearing is obtained. When the cock has been ground tight, the nut on the bottom of the plug should be adjusted so that the latter will turn fairly easily. Never hammer or batter a stopcock.

STEAM TRACTION ENGINEERING

Fitting Throttle and Gate Valves.—Throttle valves are usually of two distinct types, namely the wedge disk and the roller or butterfly. The wedge disk throttle and gate valve are almost identical in construction and operation and can only be repaired to a limited extent. Sometimes a small copper washer placed in the socket between the disk will make the valve tight again by causing the disks to spread a little more and thereby fit tighter. If this does not accomplish the result, new disks will have to replace the old ones. If removable seats are used in the valve, they also should be renewed. Regrinding these valves in the field is out of the question.

The roller or butterfly throttle can usually be reground the same way as an ordinary stopcock.

A leaky whistle valve can be reground in the same manner as a check valve.

Grinding a Safety Valve.—The best way to grind a pop safety valve is with a blacksmith's drill. It must first be taken apart and thoroughly cleaned and then the valve seat covered with a coat of oil and emery dust. The valve disk should be replaced in position and placed underneath the drill spindle. With a screw-driver bit in the spindle turn the latter down until the disk is pressed against its seat snugly but not too hard. Turn the drill vigorously, being careful that the bit does not slip on the disk and also that the disk and seat are placed exactly under the drill spindle. Grind in this manner until a new bearing is obtained on both disk and seat. Be careful not to press the disk too tightly against the seat or one of the parts may spring. A brace may be used to do this job, but the method would be much slower and not as accurate as when using the drill press. After a new bearing has been formed completely around the disk and seat, put the valve together again and set it to the required pressure carried. Instructions for set-

THE REPAIRING OF BOILERS

ting a pop safety valve will be found on the next page.

Some of these valves have a flat seat, while others are provided with a tapering one. Either type may be reground in the same manner.

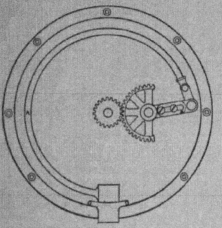

FIG. 33.—VIEW SHOWING METHOD OF ADJUSTMENT OF STEAM GAUGE.

Testing and Adjusting Steam Gauges.—This is a subject that requires considerable study and a great amount of care and good judgment in execution.

To test a steam gauge without the use of a regular testing machine requires that it be removed from the boiler and taken to another one which is provided with an accurate gauge in order that the two may be compared when both are attached to the same boiler. Do not draw any comparisons between the two gauges unless the gauge

STEAM TRACTION ENGINEERING

which it is going to be compared with can be vouched for as to its accuracy. Comparing two gauges both of which are inaccurate only leads to worse trouble if one of them should be adjusted to agree with the supposed accuracy of the other. Take the defective gauge to a large stationary plant—if located near one—as in the latter place there is usually to be found at least one accurate gauge in actual use. When the two gauges are compared, should the defective one register too much, it will indicate that the spring is weak, and if it registers less than the right amount, the spring is too strong. The inside construction of a single spring gauge is illustrated in Fig. 33, where A is the spring and B is the adjusting lever. This adjusting lever is simply a small slide held with one or two small screws to the segment gear which turns the pointer on the face of the gauge. The outer end of this slide is connected to spring A by means of a small link. A close study of the cut (Fig. 33) will show that by shortening this slide the toothed end of the segment will travel farther than it did before with the same length of movement of the spring and link, while lengthening the slide will have the opposite effect.

Suppose the test gauge registers one hundred pounds while the supposedly inaccurate one only indicates eighty pounds. This would indicate that the pointer does not travel fast enough for the tension of the spring. First shut off the stopcock on the boiler and remove the gauge. After removing the pointer and dial, loosen the small screws on the adjusting slide, shorten it slightly and again tighten the screws. Replace the dial and pointer, being sure that the latter just rests against the stop pin on the dial. Put the gauge back on the boiler and turn on the steam slowly. Repeat the above process until the gauge agrees with the standard one. Remember that a very slight movement of this lever will make a considerable

THE REPAIRING OF BOILERS

difference in the working of the gauge. Should the testing gauge register more than the standard one, it shows that the pointer moves too much for the travel of the spring and therefore the slide needs to be lengthened. Proceed in the same manner as described above, except that in this case the slide should be lengthened instead of shortened. Always remember the following rule: Shorten the adjusting lever when the gauge shows too little pressure and lengthen it when too much pressure is shown. When taking the gauge apart examine all the working joints to see that they work easily and are free from dust. It is advisable to put not more than a drop of light oil on each one of the working joints and the segment teeth. The pointer should never *press* against the stop pin. Its proper position should be just to touch the pin when there is no pressure on the gauge. It should be securely fastened to the spindle or it is liable to be jarred loose and come off.

Sometimes a gauge will "stick," i.e., the pointer will move only by jerks. This may be caused by one or both of two causes. (1) A little rust or dirt has got in the joints of the gauge or the pointer is working against the dial. This can be easily remedied by removing the pointer and dial and cleaning the working parts of all dirt and rust, putting a drop of fresh oil on each of these joints and replacing the pointer and dial. In cleaning be careful not to disturb any of the adjustments of the working parts. A good way to do this cleaning is to use a small soft brush or light cloth. Do not use any fuzzy or linty material as some of it will be almost sure to attach itself to working parts. In replacing the dial and pointer, make sure that they do not touch each other. The pointer *must* be set when there is no pressure on the gauge.

(2) The other cause of the gauge sticking is that the teeth on the pinion and segment have become worn from the vibration of the gauge (Fig. 34). The worn teeth are

STEAM TRACTION ENGINEERING

always just a few in number and are usually those which are in mesh when the gauge is registering the pressure that is usually carried on the boiler. It will be noticed in this case that the gauge seldom sticks when the pressure is low, for the reason that the teeth are not nearly so much worn at this point. This can be remedied for a short time by changing slightly the position of the pinion. To do this remove the pointer and the dial, change the little pinion so that the teeth of those not worn will mesh in the worn

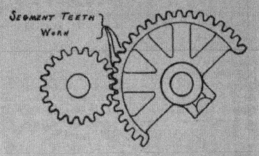

FIG. 34.—ENLARGED VIEW OF PINION AND SEGMENT OF STEAM GAUGE.

teeth of the segment. This will improve matters to some extent but eventually a new gauge will have to be purchased and the old one thrown into the junk pile.

A double spring gauge is the best type for traction or portable engines as they are less apt to vibrate and therefore will usually last longer. All steam gauges should be attached to a support or have a large connection so as to avoid as much vibration as possible. Vibrations are the hardest things that a gauge has to endure. Never close the shut off cock between the gauge and the boiler, when leaving it during cold weather, as the water in the spring

THE REPAIRING OF BOILERS

will not be drained when the steam pressure is lowered and thus prevent damage to the spring through freezing.

Another method of testing a gauge is by means of a mechanical device called a gauge tester. This is the machine by which all gauges are tested by the manufacturers leaving the factory. Certain large stationary plants possess one of these instruments, also agencies of the gauge manufacturers in the large cities and certain other people and concerns. This is by far the most accurate way to test a gauge and if it is possible to send it to a concern or person owning one of these machines for testing and correction, there will be no doubt as to its correctness upon its return.

Testing Boilers.—There are usually two methods of testing a boiler: one is known as the hammer test and the other as the hydrostatic. In the former light blows are struck all over the various parts of the shell, seams, heads, staybolts, fire box and tubes. The sounds produced upon these different parts indicate the condition of the boiler. No mere engineer can determine the condition of the boiler by this method, as it requires an expert with years of experience in this particular class of work to distinguish with any degree of accuracy between a defective and sound boiler by the mere difference in the sound of the blows of a hammer. The boiler inspectors of the insurance companies or the state and city governments are the only ones who can use this method accurately. Even these people prefer to check their reports by having a hydrostatic test made on the boiler at least once a year. The test can be made by an engineer and should be done once a year. This test should also be made after making a repair on a boiler, as has been stated before.

To make such a test, clean the boiler of all mud, scale and sediment; see that all hand holes and washout plugs are in their place. Close all outlets from the boiler except

STEAM TRACTION ENGINEERING

the one connecting the steam gauge. Remove the safety valve or whistle from the dome and fill the boiler completely full of water. The valves connecting the pipes and attachments should be made absolutely tight so that a steady pressure can be maintained. After the boiler is full of water replace the safety valve or whistle and adjust the safety valve to the testing pressure. This pressure should be fifty per cent above the highest pressure that the

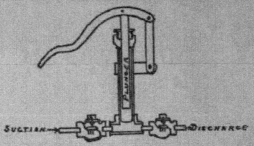

FIG. 35.—TESTING PUMP.

boiler is intended to carry. The adjustment of the valve must be done while the pressure is being increased. When the boiler has been filled and the safety valve replaced, attach a small hand force pump. With the latter, pump enough water into the boiler to bring the pressure up to the desired point. When an engine is fitted with a crosshead pump, the plunger may be disconnected from the crosshead and a handle attached so that it may be worked by hand. The Clark pump may be turned by hand by means of the small handle attached to its flywheel. Should neither of the above pumps be attached to the engine, it will then be necessary to provide a pump similar to the

THE REPAIRING OF BOILERS

one shown in Fig. 35. This pump may be provided with a one-inch pipe and half-inch check valves. Before any pressure has been put on the boiler go carefully over the fire box sides and the outer casing with a steel square or straight edge and a rule. With the square edge against the sheet carefully measure and mark the places where the sheets have been bulged, so that they can readily be referred to again. When the test is ready to be made, pump the pressure up to the desired point and have it maintained there. Have someone watch the gauge and operate the pump when necessary to maintain it, while the engineer is examining the boiler. With a light hammer tap each staybolt and closely inspect the tubes and riveted seams for leaks. While the testing pressure is still being maintained look over the bulged portions of the sheets and compare their present measurements with those taken when there was no pressure on the boiler. If the testing pressure tends to bulge the sheets enough so that it can be detected, then the boiler is not safe to carry the working pressure intended. Let the pressure slowly recede until it drops down to zero and again go over the bulged portions with the tram. By doing this it will be seen whether the sheets spring back or remain in the same position as shown by the measurements taken while under pressure. The last measurements are not necessary if no bulging is noticeable under pressure. In making a test any leaks which would naturally be expected to develop during a season's run, will appear at this time. Should leaks reveal themselves under test, do not neglect them, but determine the cause and apply the proper remedy.

Whenever a job of any consequence is done on a boiler such as putting in new tubes, staybolts or applying a patch, do not fail to test the boiler in order to determine if satisfactory work has been done. A simple way to test a boiler for leaks after putting in a new set of tubes, or after rolling the old ones, is as follows: Fill the boiler completely full of water, so that all air is excluded, by removing the safety valve. Get in the fire box and have someone light a bundle of straw placed under the fire box or the barrel of the boiler in the case it is an open bottom fire box. Of course the safety valve must have been replaced and all the valves should be tight. The heat of this small fire will warm the water several degrees, causing it to expand enough to generate the necessary pressure. As the pressure will only remain for a minute or so, it is essential that the inspection be done quickly. This test can be applied in the field, as it requires no apparatus, and may be considered a cold water test, as the temperature is only raised a few degrees. It cannot be considered of any benefit for testing the strength of a boiler because the pressure cannot be maintained or regulated.

Remember that care and good judgment must be used when testing a boiler. Never try to see how much pressure the boiler will stand, or the boiler may be weakened if not ruined. Good judgment must be used in computing the pressure to be used in the test. As stated before, the testing pressure should be fifty per cent higher than the steam pressure the boiler is expected to carry but no more. Do not make the mistake of expecting an old boiler to carry a pressure as high as, or higher than, a new one.

CHAPTER IV

THE ENGINE MECHANISM

If proper care is not taken in the operation and management of the engine, the best designed one will not be economical in the amount of steam used to obtain the required power. In order that the engineer may run his engine with the least possible amount of steam, it is desirable that he should first become familiar with the design and operation of the various parts of the machine. It is also advisable that he should know how to repair the engine if necessary. The steam engine is not complicated but it requires a nicety of adjustment and calculation to give it its best service.

Classes of Engines.—Farm engines may be divided into several different groups as follows:

According to the method of mounting
    Traction Engine
    Portable Engine
    Skid Engine

According to the style of boiler used
    Vertical Boiler
    Return Flue Boiler
    Locomotive Boiler

According to the number and arrangement of cylinders
    Simple single cylinder
    Simple double cylinder
    Tandem compound cylinders
    Cross compound cylinders

In all of the above classes, the same general principles

STEAM TRACTION ENGINEERING

are maintained, hence a description and study of one will suffice for all.

The Cylinder and Steam Chest.—The cylinder (Fig. 36) is the part in which the steam is converted into power, by imparting a reciprocating motion to the piston contained within it. It is made slightly longer than the stroke of the

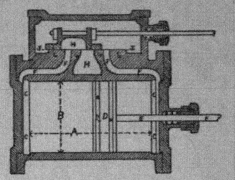

FIG. 36.—ENGINE CYLINDER AND VALVE.

engine to prevent the piston from striking the cylinder heads. This extra space at each end of the cylinder is called the "clearance."

In the cut A is the stroke of the piston, B the bore or diameter of the cylinder, CC the counterbore, D the piston, E the piston rod, FF the steam ports, GG the bridges, H the exhaust port, I the valve, J the valve seat and K the piston rings. The piston is usually a hollow casting which is turned a few hundredths of an inch smaller than the diameter of the cylinder. This allows the piston to work freely in the cylinder. It is fitted with either two or three piston rings made of cast iron or other metal.

THE ENGINE MECHANISM

These rings, which fit in grooves turned in the bearing surface of the piston, are intended to make the piston steam-tight. For this reason they are non-continuous, having a break at some point, and are turned up slightly larger than the cylinder bore. They require therefore to be sprung in place and their elasticity causes them to press against the cylinder walls, thus making steam-tight joint. As these rings continue to wear they spring outward, thus taking care automatically of all wear. Eventually they become so worn that no spring will remain in them, at which time it will be necessary to replace them with new ones. If this is not done the piston will begin to leak.

The counterbores CC are the two ends of the cylinder which are bored slightly larger than the cylinder proper. This is done to prevent the piston wearing shoulders at each end of the cylinder. If these counterbores were left out, the piston would finally wear a slight shoulder at each end of the cylinder and whenever an adjustment of the bearings in either the crosshead or crank end of the connecting rod was made so that the location of the travel of the piston would be slightly changed, a very objectionable knock in the cylinder would result. They are bored to a depth that will allow the rings of the piston slightly to overrun the edge of the counterbore. The piston should not be allowed to overrun the counterbore sufficiently to permit the rings dropping into them, as this will result in the breakage of either the piston or the ring and perhaps both.

The length of the valve seat J is also made shorter than the length of the valve, plus its travel, in order to prevent a shoulder being worn at each end of the valve seat. Should such a shoulder be formed, the valve will be raised slightly at each end of its travel, causing it to leak, besides making it jump and pound badly.

The steam ports FF are passageways from the steam chest to each end of the cylinder. The exhaust port H has no communication with the cylinder except through the steam ports and the valve cavity in the back of the valve. The bridges GG of the valve seat are partitions between the steam ports and the exhaust port. The valve I controls the admission and the exhaust of the steam to and from the cylinder. The cut shows the plain D valve, as it is called. The operation of the engine is as follows:

The cut shows the right steam port open to the steam chest and cylinder while the hollow of the valve connects the left steam port with the exhaust port. Steam being admitted to the chest passes by the right end of the valve through the right port and tends to push the piston to the left. When the piston has been forced to the left end of the cylinder, the position of the valve is changed, being moved to the right, thus opening the left steam port between the steam chest and the cylinder. At the same time the right exhaust passages will be opened. Steam will now enter the left end of the cylinder and push the cylinder to the right. The steam contained in the right end of the cylinder will escape through the steam port to the exhaust port and pipe to the atmosphere.

It will be plain that the valve admitting and allowing the steam to exhaust from the cylinder must be set so that it will change the opening and closing of the ports at just the proper time, or else the engine will not run properly.

Engine Frame.—The engine frame may be one of several distinct types; however, only the girder type and box frame are used to any extent on farm engines. The latter is not used to as large an extent as the former except in the case of light engines. It consists of a hollow casting the length of the engine to which the cylinder is secured at one end, and the bearing for the crank shaft at the opposite end (Fig. 37).

Brackets are usually cast integrally with the frame by which it is secured to the boiler. Some builders cast them

THE ENGINE MECHANISM

separately and have them securely bolted to the frame and boiler. Sometimes the box for the main shaft is cast solid with the frame. The cylinder guides, brackets for the reverse and all other parts are bolted to it. This frame, being hollow, is almost always used as a feed water heater

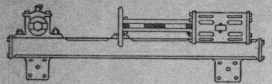

FIG. 37.—ENGINE BOX FRAME WITH LOCOMOTIVE GUIDES.

by having a coil of pipes placed inside it through which the water passes. It is heated by the exhaust steam which is exhausted into the hollow and allowed to escape therefrom by means of the exhaust pipe. This frame while being rigid and neat in appearance is open to the objection of

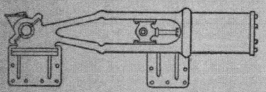

FIG. 38.—GIRDER TYPE ENGINE FRAME WITH CORLISS CROSSHEAD AND GUIDES.

expansion from the heat of the exhaust steam passing through it, and unless the valve and connecting rod bearings are adjusted while the frame is hot, they will be thrown out of their correct position by the amount of this expansion.

STEAM TRACTION ENGINEERING

The girder frame is simple, neat and strong (Fig. 38). In this style, the bored guides (Corliss pattern) are used; the guides, main shaft box and sometimes the cylinder and steam chest are all cast in one piece, which makes a very rigid construction. With this type, the cylinder is generally allowed to "overhang" the frame; i.e., it is attached in such manner that only the front head comes in contact with the engine frame. This leaves the cylinder free to contract and expand in all directions. Some builders cast

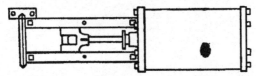

FIG. 39.—TOP VIEW LOCOMOTIVE STYLE CROSSHEAD AND GUIDES.

the brackets for attaching to the boiler solid with the frame, while others cast them separately and bolt them to the frame. The last method has its advantages, for should the frame become broken in any manner, it will be found much more easy to remove the frame from the brackets than the latter from the boiler.

Crosshead and Guides.—The crosshead is the bearing which keeps the piston rod in line, and prevents any bending or deflecting of the rod, which would be caused by the angularity of the connecting rod. It travels between bearings called guides. Both the crosshead and its guide are designed in several different styles, the locomotive and Corliss types being the most widely used. While there may be several modifications of these two types, they will always be found under the two above mentioned.

The locomotive style is usually made of four steel bars for the guides and a solid crosshead which runs between

THE ENGINE MECHANISM

these bars. A side view of this type is shown in Fig. 37 while Fig. 39 shows a top view with crosshead in position. The crosshead is not adjustable and any lost motion from wear must be taken up by changing the position of the guide bars. Side play is prevented by the guide bars having a bearing on each side of the crosshead, while the up and down motion is prevented by the top and bottom bearings of the crosshead. When these guides are adjusted in the proper manner they will do the work for which they are

FIG. 40.—ORDINARY CORLISS CROSSHEAD.

intended in a very satisfactory manner. They are open to the objection of not allowing the crosshead to adjust itself to any variation in the alignment of the crosshead pin and the crank pin and unless great care is taken they are very hard to adjust in the proper manner. These guides were almost universally used on farm engines until of late years, when they have been almost entirely supplanted by the Corliss pattern of crosshead and guides.

The Corliss type of crosshead and guide is illustrated in Fig. 38. In this type the guides are simply part of the engine frame. They are of cast iron and are cylindrical in form, being bored to a bearing which is in exact line with the bore of the cylinder. They are not adjustable, the wear being taken care of by adjustable shoes on the

STEAM TRACTION ENGINEERING

crosshead. The ordinary form of crosshead in which the shoes are removable is illustrated in Fig. 40. To take up the wear it must be set out by liners. A later and more convenient style in which the shoes are tapering and fit on a tapering or inclined face of the crosshead body is shown in Fig. 41. With this style, sliding the shoes will increase or decrease the diameter of the crosshead and thereby take up any lost motion which might occur in the crosshead and guides. The crosshead shoes are turned to the same radius as the guides, which will allow the crosshead to adjust itself to any variation of the connecting rod boxes and as the bearing surface is usually very large, the crosshead will stand hard work and long runs with very little wear or friction on them. The bearing surface of the guides is always raised slightly above the body of the casting and is made somewhat narrower than the width of the crosshead shoes, to prevent ridges being worn on the sides of the guides. They are also made slightly shorter than the travel of the crosshead allowing the shoes to run one edge slightly past

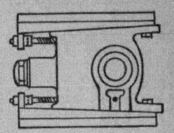

FIG. 41.—IMPROVED CORLISS CROSSHEAD.

the end of the guides. This prevents shoulders being formed at each end of the guides which would cause a very undesirable "bump" in the running of the engine.

THE ENGINE MECHANISM

Piston Rod to Crosshead.—In the locomotive crosshead (Fig. 39) the piston rod is usually fastened to the crosshead with a taper key. The rod is turned to a slight taper which fits snugly in a corresponding tapered hole in the crosshead. The tapered key when driven up tightly draws the rod very securely into the crosshead. This method has the disadvantage of not allowing any adjustment of the piston rod as to length and the clearance must be adjusted in the connecting rod boxes.

The rod of a Corliss crosshead is usually fastened by a thread and jam nut (Fig. 41). The piston rod is threaded on the end and a hole in the crosshead is threaded to fit the rod. The rod is screwed into the hole and the jam nut holds it tightly in place. The crosshead pin is fastened by means of a nut on one end. When this method is employed, the pin is turned with a large tapering shoulder on one end while the other end has one somewhat smaller than the body of the pin with the extreme outer end threaded. To prevent the pin from turning, a small pin is usually fitted in the edge of the large shoulder which fits in a recess in the crosshead.

A crosshead in which the piston rod is secured by means of a clamp is illustrated in Fig. 40. This cut shows very clearly the method of fastening the crosshead pin by means of clamps.

Connecting Rod and Boxes.—The connecting rod joins the crank shaft to the crosshead. It fits the crank pin at one end and the crosshead at the other. The length of this rod is usually made about two and three-quarters times the stroke of the engine as measured from center to center of the boxes. They are made in several different styles, of which three are shown in Fig. 42.

The old style strap and gib type, which is almost out of date at the present time, is shown in A of the cut. This rod is highly polished and usually turned and finished all

STEAM TRACTION ENGINEERING

over. The ends are rectangular in shape and slotted to receive the gib and key. The brasses are held by means of a steel stirrup fitted on the end of the rod, which is fastened by the gib and key. The gib and key are both made tapering, so that driving down the key will draw the stirrup up closer to the end of the rod, thus tightening the brasses. The key is held in position by a set screw in the side of the rod, while the gib is fastened in place by means

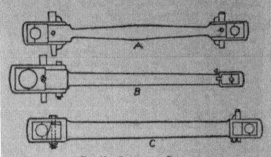

Fig. 42.—Connecting Rods.

of the lips on the ends which fit over the strap at the top and bottom. These lips also prevent the strap from "spreading." This style of connecting rod is generally used on most vertical engines and also on those equipped with the locomotive style of crosshead and guides.

A more recent style of strap connecting rod is shown as B in Fig. 42. It is being used very largely at the present time and is the only one that is adapted to center crank or double cylinder engines. These rods have the crosshead end forged solid with the rod and then slotted out to receive the boxes for the crosshead pin. The cut shows the use of

THE ENGINE MECHANISM

the key for adjusting the box. For adjusting, some makers use a tapering block with set screws instead of the tapering key. The crank end of the rod is made with a heavy steel stirrup which is carefully fitted to the rod end and secured by means of two strong bolts. The latter are turned and the holes reamed to an exact fit so that no lost motion can occur at this point. The adjustment is usually made by means of a tapering steel key held by a small set screw, although a tapering wedge is sometimes used instead of the key. When the tapering wedge is used in any form of connecting rod boxes, it is made the full width of the rod and about one-half inch shorter than the height of the slot. If a solid end rod is used, the straight side fits against the end of the rod or stirrup and the tapering side against the box. The box is also made with one side of the same taper as the block. Tap screws are used to adjust this block, one from the top and the other from the bottom of the rod.

A solid end rod used by many builders is shown as C of the same cut. It is forged from a solid piece of steel, the ends slotted out to receive the crosshead and crank pin boxes. The wedge adjustment is usually provided for taking up the wear of the crank and crosshead pin boxes. This rod cannot be used on an engine of the center crank type, on account of not being able to get the rod connected to the crank pin; however, should it have a strap on the crank end, it can be used also on a center crank engine. The connecting rod boxes are usually made of brass divided into halves, so that any wear of the boxes and pins may be easily taken up. A few makers line these boxes with high-grade babbitt metal. In some cases where the boxes are lined in this way, they are made of cast iron.

Crank Shaft.—The crank shaft is generally made of steel, but a few builders use a very high grade of hammered wrought iron. For a single cylinder side crank engine it

STEAM TRACTION ENGINEERING

is a straight shaft with a disk securely fastened on one end in which the crank pin is placed. The usual method of putting these disks on is to have the end of the shaft turned to a very slight taper and the key seated the length of the hub of the crank disk. The hole in the disk is reamed to the same taper (only slightly smaller) and key seated. The disk is then forced on the shaft by a powerful press and a tight-fitting key driven securely up and then cut off. This way of securing the disk to the shaft is a very substantial and satisfactory method and the disk will seldom work loose. The crank pin is usually set in the disk in the same manner, only instead of a key, the end of the pin is riveted down. Some builders instead of riveting it use a nut on the end of the pin. With either single or double center crank engines a different method has to be employed. The method of making these shafts is to forge the complete shaft and crank or cranks in one piece without welds and then slot out the crank pins, the whole shaft and crank pins being turned and finished in a lathe. The crank arms are planed and the balance weights are securely fastened to them. These weights are placed opposite the crank pins. They are intended to counterbalance the weight of the connecting rod, crosshead and piston. If they were not used, the engine would shake destructively every time the rapidly moving reciprocating parts were stopped or started. As these balance weights are placed opposite the crank pin, they always travel in the opposite direction from the crank and connections and if properly proportioned they will absorb nearly all of the vibrations of the moving parts. In some engines of the center crank style the balance weights are cast with a rim, which, when attached, makes the crank almost identical with the plain disk crank. In some crank engines the balance weight is cast in the disk opposite the crank arm. The different types of shafts are shown in Fig. 68. A is a single side crank shaft; B is a

THE ENGINE MECHANISM

double cylinder engine crank shaft and C shows one of the "Disk" style of balance weights detached.

Main Bearing.—These are the bearings which carry the crank shaft and therefore are made on a very heavy and substantial pattern. There are two different styles of bearings both of which are illustrated in Fig. 44. The "quarter box" A is not used to very great extent on modern engines unless they should be of small power and light construction. They have been succeeded by what is known as the "half-

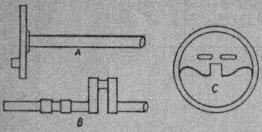

FIG. 43.—CRANK SHAFTS.

box" B or forty-five degree box, so called because they are split at an angle of forty-five degrees. The quarter box consists of a casting with large recess in it which holds the bearing composed of four pieces. These pieces are usually made of brass and are adjustable. In place of brass some builders use cast-iron gibs lined with babbitt. The lower part may be raised and thin shims placed under it, while the top quarter may be set down by means of the cap over the box. The front and back quarters are adjustable by means of set screws which screw the box proper and set the quarters closer or farther away from the shaft, whichever may be desired by the operator. The cap is held in place by either two or four bolts. These boxes are capable

STEAM TRACTION ENGINEERING

of very fine adjustment but are open to the objection of being apt to get the engine out of line unless great care is taken when adjusting them.

The half box bearings B, Fig. 44, are cast in two pieces, the base and lower half being in one piece and the cap in another. The dividing line is at an angle of forty-five degrees so that the wear will not come at the joints of the bearings but instead will be on the cap and base. The bore of the box is cast somewhat larger than the shaft and the space thus provided is filled with high-grade babbitt metal.

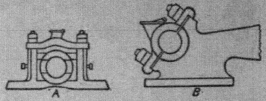

FIG. 44.—MAIN SHAFT BOXES.

Cardboard or thin tin liners are placed between the halves when it leaves the factory and these may be gradually removed and the cap set down as the bearing wears, thus taking care of all lost motion. On a large number of engines using this style of box, the lower half is cast with the engine frame. All Corliss pattern engines are built in this way. Some builders use this bearing in its plain form, lubricating the shaft by means of a groove cast on the top half which has several small holes drilled through the cap and babbitt and communicating with grooves cut in the babbitt. Others cast also an oil chamber in the bottom and still others use a small chain or ring running around the shaft in a groove cast in the babbitt lining of the box. The cap is held in place with from two to six bolts. Some

THE ENGINE MECHANISM

engines are fitted with the quarter box on both ends of the main shaft, others have it on the crank end only and still others have it on the flywheel end only. The half box may be either used on both ends of the shaft or only on one end.

Eccentric.—The eccentric which is used to drive the valve consists of a circular disk 1 (Fig. 45) which is secured to the shaft and revolves with it. In the cut 3 is the

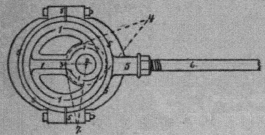

FIG. 45.—ECCENTRIC.

center of the disk and 2 the center of the shaft. As the shaft revolves the center 3 of the eccentric will describe the dotted circle. The eccentric strap 5 and the eccentric rod 6 to which it is attached will be moved horizontally during a half revolution a distance equal to the diameter 4 of the dotted circle. The distance 4 is called the travel of the eccentric while the distance 2 to 3 is spoken of as the throw. The travel is always twice the radius. It is plain that the eccentric is equal to a crank the length of which is the same as the radius of the eccentric. If the end of the eccentric rod were attached to a crank of this length, the latter would give the same motion to the rod that is imparted to it by the eccentric. The connection between the eccentric rod and the valve stem is accomplished in a

STEAM TRACTION ENGINEERING

variety of ways. Some engines use a rocker arm for this purpose, while others use a small slide or guide.

Flywheel and Friction Clutch.—The flywheel is a very important adjunct to the engine, as without it its motion would be very unsteady. In fact, without one, a single cylinder engine would not run at all and a double engine would only run with a very jerky motion. The flywheel is

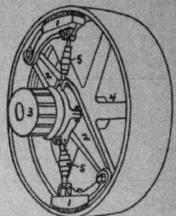

FIG. 46.—FLYWHEEL AND FRICTION CLUTCH.

almost universally used as the belt wheel. It should be accurately balanced and of sufficient weight to make the engine run steadily. A single cylinder engine requires a heavier flywheel than a double one. The rim should be turned up smoothly with the center of the face slightly higher than the edges of the wheel, so that the belt may run true on it. When the friction clutch is inside of the flywheel, as is the case with most engines, the inside is also

THE ENGINE MECHANISM

turned up true. In some engines of the center crank type there is a special pulley for the clutch which is attached on the opposite end of the shaft from the flywheel. The flywheel is secured to the main shaft by means of a large heavy tapering key which is prevented from slipping by a set screw.

The friction clutch is a device for engaging or releasing the engine from the traction gearing. This device is likely to be the cause of more trouble than any other attachment on the engine, because it requires a nicety of adjustment not easily obtained without diligent care and patience. A standard form of friction clutch is illustrated in Fig. 46 in which are the movable shoes which engage the rim of the flywheel, 2 is the spider which carries the shoes and revolves loosely on the main shaft 4. The spider has a long sleeve to which the main pinion 3 is securely fastened. The pinion engages with the traction gearing. The adjustable toggle links 5 connect the movable shoes to the movable collar on the spider hub. A trunnion strap encircles this collar which is attached to the engaging fork. This fork is connected by a rod to a lever near the foot board. By moving this lever the collar is shifted endwise on the spider hub, thereby straightening the toggle levers and causing the shoes to engage the flywheel rim. The toggle levers are provided with turn buckles and lock nuts to give a means of quickly adjusting the shoes. A turn of these turn buckles in either direction may make the clutch work very nicely or very unsatisfactorily. The amount of clearance which should be allowed between the friction shoes and the wheel when the clutch is thrown out, in order that it will hold sufficiently when thrown in, can only be obtained by trial and experience. If it does not hold or if it sticks, it is due to improper adjustment, not to the fault of the clutch or the manufacturer. All joints of the clutch except the friction shoes should be kept well oiled. This is esp... true of the trunnion on the collar and the spider where the shaft passes through it. A great occasion of annoyance with the clutch is caused by the spider being worn on the shaft, making the clutch wobble. If this joint is kept well oiled when the clutch is not in use—i.e., the engine is not delivering belt power—very little wear will occur and this trouble will be largely avoided. Nearly all up-to-date clutches are so constructed that the toggle links slightly pass the center when the clutch is thrown in, making it self-locking. When a load is going to be pulled over a steep hill, be sure that it is in good shape, so that there will be no danger of its slipping. If there is any doubt about the clutch holding, put in the tight gear pin. If the throttle is handled properly just as big a load can be pulled by using the pin as can be done with the clutch, and without danger of damage in the way of torn gearing or a bracket pulled off the boiler when using the pin instead of the clutch. Although the clutch is a very handy and practical device, it is also a source of much abuse by a great many engineers. An operator should never suddenly throw in the clutch when the engine has obtained full speed, in order to "jump" over an obstruction in the road. This sudden application of power to the gearing puts an enormous strain on the entire engine.

Pin Clutch.—Nearly all engines fitted with a friction clutch have also a tight gear pin or pin clutch. The latter consists of a round or square pin fitted in a hole in the hub of the flywheel which when pushed inward fits in a recess in the clutch spider. The pin is usually held in either position by a set screw to prevent its slipping in or out of engagement. Always be sure that this set screw is tight against the pin when in either position, for if it is not, the pin may slip out when traveling up a hill and cause some serious damage. Should the pin slip in because of improper fastening and engage the gearing when

THE ENGINE MECHANISM

the engine is doing belt work, the consequences can well be imagined. An outline of a pin clutch of late design is shown in Fig. 47. In this cut 1 is the shaft; 2, the main pinion; 3, the clutch housing which also forms a small belt pulley; 4, the clutch pin, and 5, the pin holder. The pinion has holes around its edge in which the pin fits when the clutch is engaged. The pin has a small crank on its

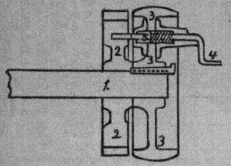

FIG. 47.—PIN CLUTCH.

outer edge, the turning of which throws it in and out of gear. This is accomplished by means of the pin holder being formed with a spiral on its outer end and on which the edge of the crank works. A spring pushes the pin in gear and holds it tightly in its two positions without slipping. This clutch needs no adjusting, is not subject to breakage or wear and requires very little attention to keep it in good working order. See to it that the housing does not slip on the shaft and also that the pin and spring are given occasionally a few drops of oil. For heavy traction work or when using the engine for long periods at road work, such as plowing, hauling, road grading,

STEAM TRACTION ENGINEERING

etc., it is advisable to use the pin even if the engine is fitted with a friction clutch. By doing so possible wear on the friction clutch will be removed and any end thrust on the main shaft, which might be caused by the clutch, will be prevented, especially if it is not adjusted properly. There are a number of different makes of pin and friction clutches on the market but they all have the same general features of action as the ones described above.

Throttle.—The three distinct types of throttle valves which are in the most common use today are illustrated in Fig. 48; while these types may vary with different

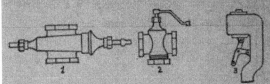

FIG. 48.—THROTTLES.

makes, the general principles of construction and operation remain the same as those described below.

The wedge disk, or slide throttle (1 in the cut) consists of nearly a straight passage for the steam with a recess at right angles with this passage. On each side of this recess and surrounding the steam passages are fitted two circular seats set at a slight angle with each other. A wedge of the same angle as the incline of the valve seats is fitted with a circular disk on each side of it. These disks have a ball and socket bearing in their center. A stem is attached to this wedge which extends outside of the valve body and is used to actuate the valve. Moving this stem back and forth opens and closes the valve in proportion to the amount the stem is moved.

The butterfly type (2) of throttle is simply a balanced

THE ENGINE MECHANISM

steam valve arranged with an outside handle for operating. A quarter turn of the operating handle fully opens or closes the valve. They are usually very durable and are easy to open and close. Both of the above types of valves if properly packed and handled will stand in any position without being fastened.

The locomotive or poppet type (3) of throttle gives a very quick opening and is not subject to wear to any great extent.

Some means must be provided, however, to hold this type of valve in place, which is usually accomplished by means of a notched quadrant for the operating lever.

The throttle is usually placed between the governor and the boiler and is operated either by a handle directly attached or a rod and conveniently located lever if the valve is out of reach from the platform. The lever is usually mounted on a quadrant. In order to become an expert in the handling of a traction engine, it is necessary that the engineer know how to operate the throttle properly without wrecking the engine or himself. In starting an engine *never* pull the valve wide open at first, for if this should be done the engine may be damaged. Instead, open it slightly, just enough to admit a sufficient quantity of steam to keep the engine moving slowly for a few revolutions. This will give the water of condensation time to be discharged from the cylinder. Now proceed to open the valve slowly until the engine has arrived at its normal speed. It requires long practice with the use of this valve in order to be able to start the engine without any jerking and in the shortest possible time under all conditions of work. Another thing which the engineer has to learn from experience is the amount of steam required to do a certain piece of work. As an instance of how to operate an engine in the proper way, take the case of one ready to be coupled to a separator located a few feet away.

The wheel of the engine being in such a position that there is no need to touch the steering wheel, should the engine be in the wrong quarter to back up, take hold of the throttle with one hand and the reverse lever with the other. Throw the reverse lever over, give the throttle a quick open and shut movement, just enough to move the piston, and with a quick movement throw the reverse lever back before the crank pin passes the center. Now open the throttle just enough to keep the engine moving in the direction wanted. Only admit enough steam to move the engine slowly and keep it under perfect control. This is a much better way of handling than to come back with a rush, perhaps smashing someone's fingers or losing control of the engine and smashing something—perhaps yourself.

Cylinder Cocks.—The cylinder cocks are small lever cocks screwed in each end of the cylinder, and connected by a lever or reach rod to the footboard of the engine. Their purpose is to drain the cylinder of condensed water when starting, to relieve it in case of foaming or priming and to drain it in freezing weather. They should be opened when the engine is shut down and not closed again until dry steam issues from them after the engine is started. Do not leave them open while running unless in the case of priming, as they simply waste that much steam and also make a mess around the engine.

Displacement Lubricators.—The lubricators used on traction engines are manufactured in a number of different designs and of two distinct types. The types are known as the water displacement and the hydrostatic lubricator. The inside construction of one of the former is shown in Fig. 49. It will be noticed that a small pipe extends from the outlet at the bottom of the cup nearly to the top. The stem on the bottom of the cup screws into a pipe or into the steam chest direct and is

THE ENGINE MECHANISM

provided with a valve so that the outlet from the cup to the steam pipe or chest can be opened or closed at will. A small drain cock is placed in the bottom of the cup for draining off the water when refilling or to prevent freezing. In starting one of these cups, first open the little drain cock and blow out all the water, close the valve on the connecting stem and remove the filler plug on the

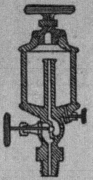

FIG. 49.—WATER DISPLACEMENT LUBRICATOR.

top of the cup. Now close the drain cock and fill the cup with cylinder oil. Replace the filler plug and open the large valve again. It is best to open this valve only slightly as otherwise the oil will be fed too fast. Steam will enter the cup through this valve and the inside tube. It will be condensed by coming in contact with the oil and since water is heavier than oil, it will sink to the bottom of the cup. The condensed water will displace the oil upwards, forcing it out through the tube and into the steam pipe or chest. The water displacing the oil causes the cup to feed.

STEAM TRACTION ENGINEERING

Hydrostatic Lubricators.—The water of condensation in the condenser and its pipe being elevated above the oil chamber, forces the oil out of the latter by just the amount

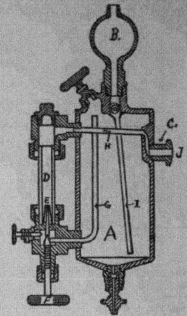

FIG. 50.—SECTIONAL VIEW HYDROSTATIC LUBRICATOR.

of pressure caused by this water above the oil outlet or drop nipple. The higher this volume of water the more positive the oil feed. As soon as the oil drop leaves the nipple, it ceases to be actuated by this pressure and rises

THE ENGINE MECHANISM

through the water in the sight glass merely by the difference between its specific gravity as compared to that of water. From the upper arm of the sight glass the oil passes into the oil tube and thence to the part being lubricated. The inside of a sight feed lubricator is shown in Fig. 50, in which A is the oil chamber; B, the condenser;

FIG. 51.—DOUBLE CONNECTION LUBRICATOR.

C, the supporting arm and oil outlet; D, the sight feed glass; E, the drop nipple; F, the regulating valve; G, the oil tube to drop nipple; H, oil outlet tube, and I, the water tube. Sight feed lubricators are made in a large number of styles but all use the same general principles of feeding the oil as shown in the cut. They are made with either single or double connection. A double connection lubricator is illustrated in Fig. 51, while the single connection one is shown in Fig. 52. The double connection style is usually much more reliable and satisfactory in its operation than the other style of lubricator. Single connection lubricators, however, are easy to attach and if properly handled will generally work satisfactorily. They

FIG. 52.—SINGLE CONNECTION LUBRICATOR.

may be attached to the steam pipe or directly to the steam chest.

The double connection lubricator is used almost exclusively on stationary engines and is not used to any great extent on traction and portable engines, because of the small amount of room for attaching it to the steam pipe, as both connections must be made on the same side of the throttle or governor. *Never* attach one connection below and the other above the governor or one on each side of

THE ENGINE MECHANISM

the throttle, as trouble will be sure to result from such method of attachment. The lubricator should be attached to the steam pipe above the throttle, as then the throttle and governor are both lubricated. A double connection lubricator may be attached to a single opening in the steam pipe, as shown in Fig. 53. The length of the nipple between the steam pipe and the tee should be as short as pos-

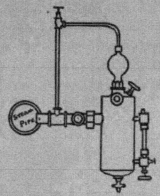

FIG. 53.—METHOD OF ATTACHING DOUBLE CONNECTION LUBRICATOR TO ONE OPENING IN STEAM PIPE.

sible to prevent too much strain on the connections. The loop from the tee to the upper connection of the lubricator is best when made with a height of ten to twelve inches above the lubricator. This gives a better water pressure within the lubricator and therefore a more forcible feed. On engines where the cylinder or steam pipe is well toward the front end and away from the platform, attaching the lubricator to the steam pipe directly places it in an awkward position, as it cannot be filled without leaving the

STEAM TRACTION ENGINEERING

platform and mounting the front end of the engine. Neither can the feed be changed or distinctly seen without doing the same thing. In order to overcome this disadvantage replace the nipple between the steam pipe and the tee by a piece of pipe long enough to place the lubricator close to the platform. A brace must be provided to steady the pipe and lubricator when attached in this manner. Always have all pipe joints perfectly tight on all lubricator connections. As long as the lubricator stays warm it will work at almost any distance from the steam pipe.

After attaching a lubricator, open all valves and drain systems and allow live steam to blow through for several minutes. This will insure all openings being clean and free from dirt. Then follow the directions given with all lubricators for filling and regulation.

On account of the many different styles and makes it would be almost impossible to give a set of instructions for handling all kinds. Be particular to get the lubricator attached so that it will stand plumb in order that the oil drop will pass up through the right feed glass without touching the sides. Always keep the drop nipple clean and be sure to drain the lubricator in freezing weather. Use only good oil, as cheap oil will clog the lubricator, blur the sight glass and choke the oil ducts. Cheap oil will also leave a refuse which if allowed to accumulate in the cylinder and rings sufficiently will be the cause of considerable damage and loss of power. This alone will far exceed the difference in cost between good and poor oil. Remember that a small obstruction in any of the passages will easily stop a cup from working; therefore always use clean oil that has already been strained. This may seem troublesome at first, but the trouble which will be saved will more than be repaid. Sometimes the glass becomes dirty and the oil drops adhere to it. To

THE ENGINE MECHANISM

clean it remove the cap over the glass and with a soft rag or piece of waste wrapped around a stick, clean the glass and nipple. Next put about a half-teaspoonful of common salt in the glass, replace the cap and allow time for the salt to dissolve before starting to feed. Always keep the lubricator in working condition and see that it is working when the engine is running. Oil is an essential thing in the cylinder in order that the engine may run smoothly as well as give full power and be durable. Three drops a minute is sufficient oil for the ordinary engine under all ordinary conditions. The hydrostatic or the water displacement lubricator is used only on the steam pipe between the boiler and the engine for the lubrication of the cylinder. The other parts on the engine which need to be fed with oil have sight feed oilers of a different design.

Oil Pumps.—Oil pumps are rapidly replacing the lubricator for cylinder lubrication. They are usually driven from the valve stem or eccentric rod by a small driving rod having suitable connections. Being attached in this way, they run only when the engine is in operation and therefore save a quantity of oil which might be wasted with a lubricator. There are a large number of oil pumps on the market, embodying quite a number of different mechanical features, but nearly all working on the same principle as an ordinary force pump. Some use a slotted plunger and thereby abolish the valves, while others use small check valves similar to a water pump. The oil pumps may either be fitted with or without a sight feed attachment, but all are so arranged that once the feeds have been fixed little attention need be given to them except to keep them filled with oil. The methods of regulating or changing the oil feed are almost always of a different construction with each separate make of pump so that no instructions of this nature can be given which would cover all makes. However, the following few hints and suggestions on the care of oil pumps in general may help to avoid trouble with them.

Always use clean oil in the pump. If the pump has any packed joints see that they are in perfect shape and do not leak. In cold weather it is best to have the pump kept warm either by a heating chamber in the pump or by having it secured to the steam chest of the engine. Look after the driving connections and see that they are in shape. Loose driving connections on a pump will make considerable noise and they impart an uneven motion to the pump. The driving pawls should be examined often, to make sure that they do not slip or allow the ratchet wheel to "backlash," as this causes a very uneven supply of oil to the engine. In attaching an oil pump, the check valve on oil pipe from pump should be placed as close to the steam pipe as possible. Try to have the oil discharged into the steam pipe *above* the throttle so that it and the governor valve may both be oiled. An engine of ordinary size usually only needs one to one and a half pints of cylinder oil during the course of a day's run. More oil will be required with a large engine doing heavy traction work, when the water is bad or when the boiler is foaming or priming.

Oil and Grease Cups.—Oil and grease cups for contact and rubbing surfaces are made in so many different styles and shapes and have so many different ways of setting and attaching them, that only a general line of suggestions can be given. Oil cups for soft oil are in two separate classes, each divided again several times. There are two styles in the first class or plain oil cups. The first style is an oil reservoir with either a screw or hinge top. Oil passes from the cup to the bearing through a small hole in the bottom of the cup. Waste or wick should be placed in the bottom of the cup to prevent the oil from running out of the cup too fast, otherwise no regulation of the oil feed is provided. The other style of plain oil

THE ENGINE MECHANISM

cup is what is called a siphon cup. A small tube extends upward from the oil outlet to a point near the top of the cup. A small wick is placed in this tube and hangs over the top edge of the tube down into the body of the cup. The friction of the moving bearing draws the oil through this wick in the same manner that the wick of an oil lamp draws its supply of oil through capillary action. The outside view of a plain oil cup of either the plain or

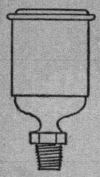

FIG. 54.—PLAIN OIL CUP.

siphon feed style is illustrated in Fig. 54. The difference between these two styles is in their inside construction.

The second class of oil cups are those of the glass body drop feed type. They may be fitted with either a plain feed or a sight feed feature. The body of the cups being of glass, a glance will indicate whether the cup is filled with oil or not. Many different ways of regulating the feed have been designed but the usual principle is to employ a small feed stem which is adjusted from the top of the cup. The stem opens or closes the orifice or outlet

STEAM TRACTION ENGINEERING

at the bottom of the cup. By this means the amount of feed to the bearing can be regulated to any quantity desired. Some styles of cups which have a small glass tube fitted in the shank or attaching arm are provided with a drop nipple and therefore the feed can be plainly seen and regulated to a drop. A glance at a sight feed cup will tell whether the cup is working and also at what rate it is

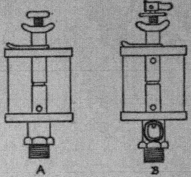

FIG. 55.—GLASS OIL CUPS.

feeding the oil. In Fig. 55, A shows a standard make of glass oil cup without the sight feed features and B shows the same style of cup with the sight feed. This class of oil cup is adapted to any stationary bearing or those with a slight motion, such as guides, main bearings, eccentrics and valve slides. For direct attachment of cups to crosshead or crank pin bearings a special cup should be provided. For this kind of bearing, also for eccentrics and main bearings, hard oil or grease cups are preferable and are also much cleaner and easier to handle than the fluid oilers.

THE ENGINE MECHANISM

The automatic or hand feed grease cups, like the oil cups, are made in a large number of different styles. The automatic type is used on crank pins, although they may be fitted to any kind of a bearing. A common type of automatic is the spring compression style, the inside construction of which is illustrated in Fig. 56. The feed is regulated by the small screw A in the shank of the cup. The

FIG. 56.—SECTIONAL VIEW OF SPRING COMPRESSION GREASE CUP.

pressure which causes the cup to feed is obtained by means of the packed plunger and the coiled spring. Another style of crank pin grease cup is made by having a cup body with a small hole in the shank in which a small three-cornered copper plunger works. This plunger should have about three-eighths of an inch play between the tip of the cup and the crank pin. The motion of the cup when the engine is running, throws this pin up and down, thus carrying the grease to the pin at every stroke of the engine. This cup cannot be used on a stationary bearing, as a quick motion is required to make the cup feed. For stationary bearings, the plain screw compression cup is well adapted, being cheap, simple and durable. Keep the grease and oil that is used in the cup free from dirt and deposits.

Governor.—The governor is to the engine what the brain is to a man. When an engine is running at a uniform speed the work done in the cylinder must just equal the resistance to be overcome at the rim of the flywheel plus the frictional resistance of the engine. Should the resistance become less than the work, the excess of work would cause the moving parts to move faster and faster and the engine would race or run away. But should the resistance exceed the work, the engine will slow down until it would finally stop. The work required of the engine cannot always remain constant and therefore it is necessary that some means be provided to adjust automatically the steam supply to the variation of the resistance. This is the purpose of the governor.

Governors may be divided into two distinct classes: (1) throttling governors which throttle the steam in the supply pipe, and (2) automatic or adjustable cut-off governors which regulate the steam supply by changing the point of cut-off. The automatic governor is often a shaft governor varying the cut-off by changing the radius of the eccentric which in turn changes the length of the valve travel, giving a long or short cut-off according to the load on the engine. This governor is not adaptable to farm engines. The Corliss valve gear and governor is another one of the automatic type through the medium of changing the cut-off, but it is not adaptable to farm engines.

The throttling governor is used exclusively on traction engines. As stated before, this class of governor regulates the speed of the engine by changing the weight of steam passing through the steam pipe. An outline drawing of the well-known Pickering governor is illustrated in Fig. 57. The three balls, B, are fastened on the three flat springs, S, the latter being in turn fastened to the top and bottom collar, E. All this constitutes what is called the

THE ENGINE MECHANISM

revolving head. This head is given motion from the crank shaft of the engine by means of a belt which runs on a

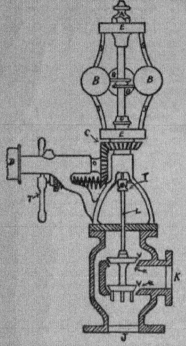

FIG. 57.—GOVERNOR.

pulley on the crank shaft and the pulley D, on the driving shaft of the governor, and thence through the miter gears,

STEAM TRACTION ENGINEERING

C, one of which is secured to the drive shaft and the other to the bottom ring of the head. Steam enters the governor at K, passes through the valve seat, V, as indicated by the arrows, and into the engine through the outlet, J. The valve V is a double seat valve and since steam is on both sides, it is balanced and moves very easily. The valve stem, I, passes upward and has a bearing at A of the revolving head. The valve is held open by the tension of the speeder spring and the lever, T. The centrifugal force of the revolving balls tends to force them further from the axis which pulls down the upper head. This presses down the valve stem and closes the valve, shutting off the supply of steam. When the load on the engine is light the engine will tend to speed up. This speed is imparted to the revolving balls, forcing them farther away and closing the valve more and more. When the speed drops, the balls return toward the center, thereby opening the valve and causing more steam to be admitted. Different speeds may be obtained by changing the tension of the speeder spring, T. This changes the resistance to the centrifugal action of the revolving balls and by this means changes the speed of the engine. For more speed the tension of the spring is increased and for less speed the tension is lowered. The above description will answer for all throttling governors as to general principles. The design and manner of changing the speed varies with the different makes. A number of the latest designs have a speed changing device which does not make it necessary to change the spring tension to change the speed. These designs accomplish it by changing the length of the valve stem. The recent Pickering model and a new Judson are of this construction. The Eclipse, Monarch and Gardner governors vary the speed by a speeder spring. A form of governor used on the Reeves engine has a rather unique method of changing the speed without changing the spring tension.

THE ENGINE MECHANISM

It is accomplished by changing the position of the fly-balls on the revolving head. In other words, when the balls are thrown farther out on the head the speed will be decreased, while bringing them in will have just the opposite effect. The adjustment of the speeder spring is usually made by means of a thumb nut and a small worm gear. The quick action speed changes on the others are adjusted in different ways according to the make of the governor but always through the medium of a small wheel or handle placed in some convenient position to the operator. The Waters governor is adjusted by screwing the valve stem up or down (up for more speed and down for less), a thumb nut being secured to the top of the valve stem for this purpose. The stem is held in position from turning out of adjustment by a jam nut.

Having observed how the governor is constructed and how it is regulated, it is necessary to determine what might happen to prevent it from properly performing its duty. Suppose an engine runs with a very jerky motion, first almost stopping and then all at once seeming to take a notion to race. When this happens stay close to the throttle and watch the governor closely. It will probably be noticed that when the engine goes off on one of its races the stem will suddenly go down through the stuffing box and stay there until the engine slows up again, when all at once up goes the stem and away goes the engine on another race. The trouble usually is that the packing has become old and dry or is screwed down too tight. To remedy the trouble loosen the packing nut until it leaks a little steam and apply plenty of oil. The stem should be repacked at the first opportunity. The trouble may be due to a bent or rusted or scored stem, binding in the stuffing box. If this is the trouble the best remedy is a new stem. It is rather a particular job to straighten a bent governor stem, or to pack one whose smooth surface

STEAM TRACTION ENGINEERING

is gone. It is best not to try. Sometimes a new governor will cause the engine to act this way. When it does the trouble is very likely to be in the valve. It may have a rough place or a small projection on it somewhere which does not allow the valve free action. Take the valve out and use a little emery paper on it for a few minutes. After this has been done the governor should operate properly. If the engine has been run any length of time and still acts this way, the trouble is very seldom in the valve. In this case first examine the belt to see that it is not slipping. If the belt is all right the engine should be stopped and the governor examined to see if its pulley is not a little loose on its shaft, causing it to slip. If the trouble is due to neither of the above causes the governor pulley on the crank shaft should be carefully examined, as it is very apt to become loose. A few drops of oil on the joints of the governor will often stop this racing, since when the governor bearings get dry considerable more power is consumed in driving it and this resistance may make the belt slip. Sometimes with old engines, the gears on the governor become so badly worn that the teeth slip, causing a very irregular engine speed. The remedy is usually a new set of gears, although sometimes they may be set to mesh deeper, in which case they are apt to last a little longer.

If the engine starts off at a lively gait and continues to run still faster while the governor slackens its speed, the cause of the trouble is very likely to be a loose belt. Should this be all right and the speed is varied by changing the length of the valve stem, see if the jam nuts holding it in position have not worked loose. If a wheel is loose the governor will very likely stop or at least slacken its speed while the engine races. If the trouble is in the valve stem, the governor will run as usual but the engine speed will be accelerated.

THE ENGINE MECHANISM

If the engine fails to start, look to see if the governor stem is not screwed clear down or the speeder spring entirely loose, allowing the valve to drop and close the valve chamber. Should the engine suddenly stop and the main valve and governor appear to be all right, notice if steam enters the steam chest when the throttle is wide open. If no steam enters, the governor valve must have dropped down, closing the steam passage. Should the speeder spring be all right the valve may have come off of the valve stem. This valve is either secured with a nut or with a small pin, both of which may work out of place, causing the valve to drop.

Keep the governor lubricated with a good grade of thin oil. Always try to keep it clean and free from dirt, as varying resistances in its joints will make the governor action irregular. Suppose the governor belt should break or come off one or both of the pulleys while the engineer is forty or fifty feet from the engine, either nothing in the way of damage may occur at the time or else great damage may be done both to the engine and to persons in the path of the flying parts of metal. The engineer's place is at the engine and he has no business to be away from it. It takes a very short time for the flywheel to let go after the engine begins to race.

When traveling through timber be careful that the limbs or shrubs do not come in contact with the governor. A governor is not a very complicated piece of machinery, neither is it hard to handle, but it will require care and attention the same as any other machinery. When taking a governor apart be careful to lift the top off straight to prevent bending the valve stem. Use a thin belt for driving the governor and do not lace it so that it will jump when passing over the pulleys. A leather belt with small belt hooks will give the best satisfaction if properly looked after and kept reasonably tight.

STEAM TRACTION ENGINEERING

Engine Mountings.—This subject may be divided into two distinct groups which can be again subdivided:
First Group
  Independent Mounting
    a. Under mounted
    b. Top mounted
Second Group
  Boiler Mounting
    a. Side Hung
    b. Rear Hung

The manufacturers of the independent mounted engines claim that by this method of mounting they relieve the

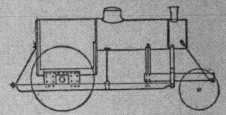

FIG. 58.—INDEPENDENT UNDER-MOUNTED MOUNTING.

boiler from the strain of the ground wheels and the traction gearing. A noted type of this mounting is illustrated in Fig. 58, which is known as the undermounted style (Avery engine). As will be seen from the cut, the entire working parts, engine, ground wheels and traction gearing are all secured to a strong steel framework. This frame takes all the stress of the engine ground wheels and gearing so that the boiler only has to carry its own weight over the engine. This method of mounting an engine is very strong and naturally makes a much heavier engine than some other styles of mounting. Another method of

THE ENGINE MECHANISM

independent mounting is shown in Fig. 59 (Aultman and Taylor); this consists of a channel steel frame to which are attached the ground wheels and the gearing. The en-

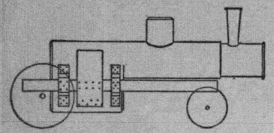

FIG. 59.—INDEPENDENT TOP-MOUNTED MOUNTING.

gine is mounted on top of the boiler in the usual way. This manner of mounting is claimed to reduce the stress on the boiler with only a very little difference in appearance from the others and with a very little increase in the weight.

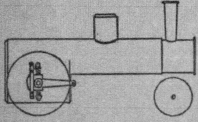

FIG. 60.—TOP-MOUNTED SIDE-HUNG ENGINE MOUNTING.

A plain side-hung boiler-mounted engine is illustrated by Fig. 60 (Advance engine). This method is used by

STEAM TRACTION ENGINEERING

several companies with very good results. The front trucks are fastened to the boiler by a pedestal, or stand, which is bolted directly to the boiler. The rear or drive wheels revolve on stub axles which are held by brackets fastened to the sides of the fire box. Some engines have these brackets mounted with springs, to relieve the boiler from road shocks, while some fasten them rigidly to the fire box sides. Both styles have their advocates and seem

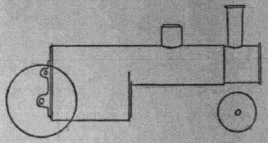

FIG. 61.—TOP-MOUNTED REAR-HUNG ENGINE MOUNTING.

to work satisfactorily. The countershaft is usually placed in brackets across the front end of the fire box and the intermediate gear on a bracket on the side of the boiler.

The top-mounted rear-hung engine is well illustrated in Fig. 61 (Rumely). In this style the front truck is mounted on a pedestal or stand in the same manner as the side hung engine, but the rear wheels are fitted on an axle which turns with them, in large boxes fastened to the rear of the boiler. This axle extends clear across the engine. The countershaft is mounted in similar boxes directly over the axle and the intermediate gear is held either by a bracket on the side of the boiler or by a shaft which runs in boxes on the top of the boiler. Some engines have the rear axle

THE ENGINE MECHANISM

and countershaft held by trunnions and radius links and connected to the boiler through springs. This construction is known as the spring mounted engine. Both styles have been used quite extensively, especially for large, heavy engines, and both have proven a satisfactory method of mounting. The manner of mounting is a matter of per-

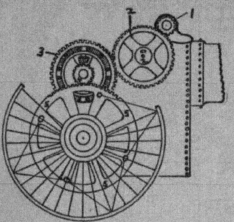

FIG. 62.—TRANSMISSION GEARING.

sonal opinion and taste both with the user and the manufacturer.

Traction Gearing.—A large proportion of the success or failure of a traction engine lies in the traction gearing. Through the medium of these toothed wheels all the power of the engine must be delivered which is required to propel the engine and draw its load. Since this is the case the gearing must receive its share of attention or it will quickly become a source of trouble, annoyance and expense. Breakdowns in the gearing are always expensive and usually happen when the engine is in such a place that it cannot be used until the damage has been repaired. This usually means a new gear or pinion.

A standard method of gearing engines is shown in Fig 62, in which 1 is the main shaft pinion, which is connected to the clutch; 2 is the idler or intermediate gear; 3, the compensating gear; 4, the bull pinion, and 5, the master gear attached to the drivers. Care should be taken to see that the gears mesh properly and that they do not become loose and work out of place or line. The bearings must receive ample lubrication and the countershaft always be kept in good condition as it carries the bull pinion and the compensating gear, which compels it to withstand a very severe strain when the engine is pulling hard. The master gears should always be kept tight and not allowed to work loose from their connection with the drive wheel. Many gear breakdowns can be traced to the neglect of this duty. If the gearing is kept in proper mesh and in line, it will give the engineer very little trouble and will give satisfactory service.

Compensating Gear.—In order that one of the driving wheels may be able to turn farther than its mate when an engine is turning a curve or corner, and at the same time compel equal driving power to be applied to each wheel, the engine has to be provided with what is known as a compensating gear. This gear, sometimes called the differential gear, is actually a combination of gears and pinions. It is usually placed on the countershaft, although there are a few builders who place it on the rear axle. Its function and operation are the same, however, in either place. Numerous styles of this gear have been designed but they all come under the head of two classes, namely, either the bevel or the spur gear types. Since both work on the

THE ENGINE MECHANISM.

same principle, the ordinary bevel type illustrated on Fig. 63 will suffice for an explanation of both types. In the cut 1 is the main spur gear which receives the direct power of the engine and revolves loosely on the hub of the bevel gear 4. This latter gear is keyed to the countershaft 3 to

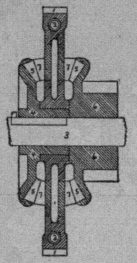

FIG. 63.—COMPENSATING GEAR.

the outer end of which is keyed one of the bull pinions. The other bevel wheel 5 has the other bull pinion 6 attached to it and both revolve loosely on the shaft. The small pinions 7 are mounted on short shafts fixed in the main gear and mesh with both bevel gears. A collar is placed on the shaft against the loose bull pinion and bevel wheel to prevent the gears from spreading and to keep them in proper mesh. The coil springs 2 are fitted to the rim of the large spur gear to relieve the gearing from heavy shocks and jars. When the engine is moving straight ahead the whole gear revolves as one piece but in turning corners the bevel pinions revolve between the two bevel wheels, allowing the traction wheels to adjust themselves to the radius of the curve and at the same time compel each wheel to do its share of work of propelling the engine. The bearing of the large spur gear where it runs on the hub of the bevel wheel should always be kept well oiled, as should the bearing of the small bevel pinion. Be sure that the bevel gear and pinions are in proper mesh and keep them so by means of the adjustable collar on the shaft. Some makers do not use springs in this gear while others do. The number of pinions varies according to the size of the engine and the special ideas of the designers. On very heavy engines as many as eight pinions may be used, but usually four or six are enough for the compensating gear on engines of moderate size. In the spur gear type of compensating gear, spur gears are used in place of the bevel gears while spur pinions replace the bevel ones. This type works in exactly the same manner as the bevel type.

CHAPTER V

CARE AND MANAGEMENT OF ENGINES

Starting an Engine.—The proper starting of an engine largely determines whether it is going to run successfully or is going to give the engineer a lot of trouble. The following few suggestions apply equally whether the engine is a new or an old one that has been idle for some time, as between one threshing season and the next.

When screwing the oil cups in place clean out the oil holes and the boxes from cinders and dirt. The cups should be screwed in firmly and filled with the proper kind of lubricant which is suitable to the particular cup being filled and the bearing which it is to lubricate. Oil all bearings and joints thoroughly some time before starting so that the oil may have a chance to soak into the bearings. Fill the recesses in the boxes with a small amount of clean waste to prevent the oil from running out too fast. Put the lubricator or oil pump and all drain cocks in position. Lace the governor belt in position and oil all joints of the governor. When there is a pressure of forty or fifty pounds of steam on the boiler, throw the reverse lever in either the backward or forward position, open the cylinder cocks and give the engine a little steam. Be sure that the lubricator or oil pump is working properly when starting the engine. If the engine is a new one it may start a little hard at first, due to tight fitting rings and pistons. There is nothing wrong about such a condition and there is no necessity for doing anything about it except to see that the cylinder and valve are kept well supplied with oil. This condition will last only a few days but during that period allow the lubricator or oil pump to feed an extra quantity of oil, after which the supply may be gradually cut down to the normal amount. A new engine which is "tight" frequently requires fifty to sixty pounds of steam to start it but in a few days it will start just as easily with fifteen to twenty pounds. Allow the engine to run slowly for a time, and meanwhile go over all bearings to see if they are heating. Should a bearing be found which is heating, stop the engine and loosen it up a trifle, giving it plenty of oil or grease, and then try it again. If it should continue to run hot, it is best to remove it and thoroughly clean it and the journal. Kerosene is one of the best materials to clean a bearing. After cleaning the bearing replace and adjust it properly (pages 225, 227), add sufficient oil and try it again. A few hours spent in this work at the start will very often save a number of hours of trouble with hot boxes later. Run the engine slowly until it is running smoothly before putting it to work. By carefully feeling the different bearings for possible hot boxes during the first few days of running, future trouble from this cause will be prevented. Sometimes the governor on a new engine will stick and fail to work. If this happens, read again the section on governors in the last chapter (page 136). If the engine refuses to start with a wide open throttle, open the pet cock on the steam chest and see if steam issues out of it. The governor stem may be either screwed clear down or the speeder spring is not holding the valve up. In either case, steam cannot enter the engine. Before starting a new engine, it should be thoroughly packed with good engine packing (page 273). For instructions in regard to the starting of a new boiler, see page 42.

Hot Boxes.—Hot boxes may be the result of the following five causes: (1) bearings too tight; (2) bearings too loose;

CARE AND MANAGEMENT OF ENGINES

(3) bearings out of line or badly cut; (4) dirt, sand or foreign substances in them; (5) insufficient lubrication. Any one of the above causes or two or more of them together will be almost sure to result in a hot box.

A bearing which is insufficiently lubricated will become hot, causing the box and the journal to expand, which will result in the journal seizing, thus increasing the heating until it eventually refuses to revolve. Sand, grit and other abrasive substances in the oil may get into a bearing and cause it to heat. Whenever a bearing which has to withstand a vibratory motion under a heavy load—the crank pin, for instance—is allowed to get excessively loose it is almost sure to get hot. If a bearing gets hot from being too loose the remedy is to see that it is properly adjusted and that it receives a sufficient supply of oil. If a nicely adjusted bearing, which previously has been running cool and has not been recently tightened, suddenly becomes hot, either it is not getting enough oil or some foreign substance has worked into the bearing. If plenty of oil does not cool the bearing, loosen it slightly, as the heat may have caused it to expand enough to become tight, and keep it well supplied with oil. Running kerosene through a bearing will often wash out any dirt; it should be followed immediately with plenty of good oil. Cylinder oil is good for a hot bearing, but better results can be obtained by using a mixture of cylinder oil and powdered flake graphite. To help cool off a bearing either sulphur or good clean salt can be used. The crank pin and main shaft boxes are usually the ones to give the most trouble from heating and therefore should be closely watched. They are liable to begin heating when least expected and for no apparent reason. When a babbitted box gets very hot it often melts the metal enough to fill the oil grooves. In this case the box should be cleaned and new grooves cut in it and when the box is torn or melted to any extent, it is better to rebabbitt it at once. When a brass box gets excessively hot, it should be removed and the box and journal re-dressed as soon as possible; otherwise it will be apt to continue to give trouble.

Lubrication of Gears.—The gearing of an engine has to withstand enormous strains, besides being subjected to a large amount of wear and tear. It works under adverse conditions in that it seldom can be effectively protected and is usually placed in such a position that it catches the sand and dirt. The proper oiling or lubrication of the gears of a traction engine has long been a problem to both the user and the manufacturer. For engines which are used for short periods on the road and are not required to do heavy hauling or plowing, fairly good results could be obtained by occasionally greasing the gears with axle grease. This method is totally inadequate for recent makes of heavy engines used for continuous heavy traction work, such as plowing and freighting. The engines of this design are fitted with some make of gear oiler. The most common form is an oil tank mounted on top of the boiler which has a number of small pipes fitted with controlling valves leading from it to the top of the different gears and allowing the oil to drip directly on the teeth while the gears are in motion. Another form is a large cup with valve and small pipe placed over each line of gearing. A cheap grade of oil known as "black jack" or "black oil" can be used in oiling the gears. Lubricating the gears with axle grease causes the latter to catch and hold sand or dirt, thereby cutting the gears nearly as fast as when no lubricant was used. Oil dripping on the teeth will also catch dirt but it will be carried away again when the oil drips to the ground, and besides the gearing will be as fully lubricated an hour after starting as when the engine was first started.

Lubrication.—A whole book could be written on this subject alone in order to cover it fully. In fact, there are a number of such books already written now on the market. A good definition of lubrication would be: The application of a substance to a bearing or surface to prevent metallic contact and thus reduce the frictional resistance between two or more bodies which would otherwise rub on each other. It has already been pointed out that a bearing will soon cut and heat if not properly lubricated. It is a well-known fact that friction produces heat and also promotes cutting and excessive wear. This frictional resistance can be reduced by the application of a sufficient supply of the right grade and kind of lubricant. The amount of lubricant must be such that the bearing, whether of the sliding or rotating type, will always have a thin elastic coating between its wearing surfaces at all times. Different kinds of bearings require different grades and kinds of lubricants. As an instance of what is meant by the above statement, it can be readily seen that a bearing on a sewing-machine will require a lighter grade of oil than the heavy bearing of a locomotive. A light grade of oil would be squeezed out of a heavy bearing in a short time. For the same reason a limpid liquid like water or gasoline would not cover a bearing for as long a period of time as an oil of proper grade.

The friction of the cylinder and valve of a farm engine is enormous; especially is this true of the valve if it is of the plain D type. Unless they are well lubricated, they will soon become badly cut and thereby ruined. Failure to use a high grade of cylinder oil by operators of farm engines has been one of the main reasons why manufacturers have discarded the old link reverse for some of the newer designs of valve gears. A valve will not run any more easily with one of these gears nor will it run with either a less quantity or a poorer grade of oil. The reason for the change is that the link will begin to jump and pound when the valve gets a little dry. The very best of cylinder oil should be used as poor oil not only allows the cylinder and valve to become cut, but it will also cut down the power of the engine to a great extent on account of the cylinder rings and valve becoming gummed up with the refuse from the poor oil, thereby increasing rather than diminishing the friction. Steam at 100 pounds pressure has a temperature about 338° F., and higher pressures have correspondingly higher temperatures. A low grade of oil breaks down and oxidizes to a gum at these high temperatures. Cylinder oil for 125 pounds steam pressure should have a fire test of about 450° F., and with higher pressures the oil should have a fire test of from 500° F. to 600° F. An oil having too high a fire test will not atomize or mix with the steam and therefore is almost as bad as one having too low a fire test temperature. A very good grade of cylinder oil is one having a fire test of 530° F., gravity of 22° Baumé, flash point of 485° F., cold test for ceasing to flow at 25° F., and a viscosity of about 130 at 210° F. The viscosity of an oil is a measure of its tendency to resist being squeezed out of a bearing under pressure. Too high a viscosity in a cylinder oil is not desirable, because it not only resists motion, but if the oil is retained too long in the hot cylinder it tends to burn and form a deposit.

Never fill up the cylinder lubricator or oil pump with bearing or machine oil and then wonder why the cylinder squeaks. These oils are too light in body and have a very low fire test. They were never intended for this kind of work. As it would be folly to try and carry several kinds of oil for all the different bearings, select one which will work satisfactorily on all of them. It is almost universal to use the same kind of oil on both the separator and the engine, and where a high grade of machine oil is used, it works satisfactorily on both. An oil giving the following tests will prove a desirable article for all around use. Gravity

CARE AND MANAGEMENT OF ENGINES

of 26° Baumé, flash of 350° F., first test of 400° F., cold test at 32° F. and viscosity of 135 at 100° F. of temperature. Machine and engine oils do not require a high fire test but they should have a high viscosity and cold test.

Hard oil or cup grease is coming into almost universal use on crank and crosshead pins, eccentric and main bearings. When used in properly constructed cups and the boxes have been properly grooved and fitted, this method is highly preferable to the old style of lubrication. Hard oil is very seldom satisfactory for the lubrication of guides or slides.

Knocks and Pounds.—A knock or a pound may arise from a number of causes; a loose box, a loose nut, a loose bracket, engine out of line, valve not properly set and a variety of other things. Loose bearings are the main cause of knocks in an engine and are sometimes very difficult to locate. First make sure in which bearing the knock occurs and also what is the reason for its occurrence. Do not try to take all the knock out of a loose bearing at one time, but instead, gradually tighten it so that it will not get hot. To the inexperienced engineer, nearly every knock in the engine will seem to proceed from the crank pin which is usually the first thing to begin knocking but it must be remembered that all knocks do not come from this source. The crosshead will knock badly if not looked after and the knock may seem to be located in the crank. The crosshead which is loose in the guides will produce a very undesirable knock and is sometimes especially hard to locate in old engines.

A knock which very often bothers young engineers is due to the piston striking the cylinder head. If the liners or shim-backing in the crosshead and crank pin boxes of the connecting rod are not evenly divided, the rod will be gradually lengthened or shortened until the piston will strike the cylinder head. To remedy this knock divide up

STEAM TRACTION ENGINEERING

the clearance of the piston at the two ends of the stroke. The best way to do this is to divide evenly the backing or shims in the connecting rod boxes. If the engine is fitted with a piston rod which is screwed into the crosshead, turn the engine on the head end center and unscrew the rod out of the crosshead until the piston strikes the cylinder head, counting the turns. Turn the engine to the other center and screw the rod into the crosshead until the piston strikes the other cylinder head. Count here also the number of turns through which the rod is screwed. Now screw the rod out again just one-half of the number of turns that it was screwed in and the clearance will thereby become equally divided.

The main shaft boxes, especially the one next to the crank, will often knock. These boxes produce more of a pound or thud than a knock. A loose box next to the crank nearly always makes a noise like a loose crank pin. Eccentric yokes will knock when allowed to get loose and give out a kind of a rattling jerky sound. In fact any bearing on the engine from the largest to the smallest will knock or pound if allowed to get loose.

Allowing the engine to get out of line will cause the engine to knock and it cannot be tightened enough to stop it without causing excessive heating or cutting. The most frequent lack of alignment is that of the crank shaft with the cylinder. Should the crosshead be improperly adjusted, the piston will not come in perfect alignment with the cylinder, which will result in cutting and pounding. Other parts about the engine which may cause knocks are: too much or too little lead on the valve; uneven cut-off, and a loose governor pulley on the main shaft. The remedy for the first two items is to reset the valve correctly. The knock of the governor pulley is generally very hard to find and is also very disagreeable.

A rattling noise in the cylinder is usually due to loose

CARE AND MANAGEMENT OF ENGINES

or worn cylinder rings or it may be caused by the valve itself. The latter happens whenever the clearance between the valve and the nut or blocks on the valve stem exceeds one-sixteenth of an inch. This clearance should not be more than one-thirty-second of an inch. No damage will result from the worn rings unless they are broken, but if the rattle should become excessive they should be removed.

Engines fitted with the Marsh reverse will often develop a very disagreeable knock which is very hard to locate. It is caused by the small crank shaft being slightly loose in the bearing, allowing the small disk to pull away and then strike against the flange on the governor pulley spool. A peculiar thing about this knock which will help to locate it is that it only appears once in each revolution, while nearly every other kind of knock will sound twice in each revolution.

Remember that trying to take a knock all at once out of a bearing will only result in causing it to become hot. When knocks result from the engine being out of line, the only remedy is to reline the engine properly. A loose bracket or wheel is just as liable to cause a knock as anything else.

Keeping Gears in Line.—Where the intermediate gear runs on a stud bearing attached to the side of the boiler, the stud and hub of the wheel will finally wear to such an extent that the wheel will wobble and be continually out of line. This will cause the gearing to bind, run hard and rapidly wear out, often resulting in the breakage of the gear. Where the intermediate gear is keyed to the shaft this trouble is partly obviated but not entirely done away with as the boxes will wear and allow the shaft to get out of line with the same result as with the stud and bracket gear. About the only remedy for the former is a new gear and if the stud is worn to any great extent it also should be replaced with a new one. In some cases the stud may be re-turned, the gear re-bored and then bushed, but this is usually an expensive job and can only be done at a well-equipped machine shop. The side play of the wheel on the stud should never be allowed to exist. Where a set collar is used on the end of the stud, the side play may easily be overcome by pushing the collar against the gear and securing it in this position. Where a cap is used on the end of the stud a washer will have to be made of the thickness and large enough to fit over the stud. On intermediate gears attached to the shaft the boxes will have to be rebabbitted when the shaft is out of line unless they are only slightly worn, in which case the trouble may be corrected by taking them up properly.

The countershaft very often gets out of a true alignment at right angles to the gear train, thus throwing the gearing out of line. When this happens it should be repaired at once. The main cause of this is either that the countershaft boxes have worn, a springing of the shaft or loose brackets. As the countershaft nearly always carries the compensating gear, its importance may be readily understood.

Sometimes the bevel wheels are allowed to spread and this will make the main gear wobble. The bevel wheels should be kept in proper mesh with the small bevel pinions. This play is usually surmounted by placing a washer between the loose bull pinion and the cap on the end of the countershaft. Where a set collar is used on the shaft opposite the keyed bevel wheel and on the other side of the box it may work loose and allow the keyed wheel to slip endwise on the shaft. The remedy is to drive the bevel wheel back into position, place the collar against the box snugly and fasten it securely. In some cases where the strain is excessive the set screws in the collar may have to be countersunk into the shaft slightly to prevent slipping. The bearings of the countershaft must always be kept well

CARE AND MANAGEMENT OF ENGINES

oiled as the wear and tear on them is enormous. The bull pinions should never be allowed to work out of line on the shaft.

Engines having the drive wheels fitted on stub axles often have the hubs of the wheels badly worn, allowing the wheels to tip, which will cause the bull gears to be thrown out of line and therefore causing undue wear on them. The hubs should be bushed so as to run true on the axles. If they are worn to a great extent it is best to have the spindles re-turned, the hubs bored and a bushing of the proper thickness fitted on the spindle. Some manufacturers use an axle with a detachable sleeve and wheels with removable bushings in the hubs. The renewing of these sleeves and bushings is a comparatively easy matter. If the drive wheels turn with the axles as in a rear mounted engine, the hubs of the wheels very seldom wear at all and therefore they will not tip or wobble but the bushings in which the axles revolve are subject to wear and this wear will throw the gears out of line. The only proper remedy for this condition is to rebabbitt the axle in its proper position. Then again the axle may be sprung in some manner and this will cause the wheels to wobble more or less. In a case of this kind either get a new axle or have the old one straightened.

The bull gears must be kept on the drive wheels or they will chatter, causing them to run hard, cut and consume a great deal of power.

Counter and Idler Shafts.—This subject has already been more fully treated in other parts of the book but a few facts concerning them which have not been touched upon will be treated here.

If the countershafts run in a solid cannon bearing (Fig. 92) they cannot be adjusted except as to being in line. This may be done by changing the radius links holding it in position. The only way to take up the lost motion is to rebabbitt the bearing (page 261). On engines where the countershaft runs in split boxes, the caps may be removed and enough of the liners taken out until the play has been removed. Never allow the countershaft box to become too tight or it will soon be ruined by the excessive pressure. In order to get the best results pay particular attention to see that the oil grooves are cut the proper amount and that oil holes are open, so that it will be able to receive its supply of oil. If the boxes are still loose after all the shims have been removed, they will have to be rebabbitted (page 246). To get the shaft in the proper place before it is removed, measure the distance between the countershaft and the axle, the bull gears being in proper mesh, and also the distance from the countershaft to the intermediate gear. In setting the shaft prior to pouring the babbitt, duplicate these measurements.

An idler shaft is taken care of in exactly the same manner and is rebabbitted the same way. The shaft is usually placed in line with the gearing by using a square in the same manner as when squaring a main shaft (page 247). Always be careful, when rebabbitting either an idler or a countershaft, not to get the gearing so close together that the teeth will bind in the bottom of the cogs as this will cause excessive wear if it does not actually result in a breakdown.

Rear Axles.—Rear axles may be divided into several different styles since they may be spring- or solid-mounted and also because they may have stub or full length axles. Stub axles have usually a large shaft cast in a bracket which is attached to the side of the boiler and are in quite common use on the lighter makes of engines. The drive wheels revolve loosely upon the axles. The bracket may be attached directly to the boiler or to another bracket through springs. This last method gives the spring-mounted style. The axle may be either solid or hollow. If they are hollow

CARE AND MANAGEMENT OF ENGINES

they are filled with waste and kept oiled in this manner. When the axle is solid and of the bracket type it is usually fitted with a removable sleeve which can be replaced when worn, thus making the axle as good as new.

Full length axles may be either solid- or spring-mounted and either turn with the wheels or have the wheels turn on them. Rear-mounted engines are always fitted with full length axles turning with the drive wheels. The bearings for full length axles may either be of the cannon or split-box type. A few builders only still use the ordinary solid box. The cannon bearing, which gives good satisfaction, is used quite extensively. When the boxes become worn they can be rebabbited in the same manner as ordinary ones (page 246). Be sure to remove the axle after rebabbitting and cut ample oil grooves and oil holes in the bearings. Always keep the bearings of the axles well oiled no matter what style is used, as they are subject to enormous pressure. Do not allow the wheels to have any end play on the axles or the axles to play endwise in the bearings. When the engine is of the spring-mounted type, frequent inspections of the springs, hangers and bolts should be made, and if any are worn very much, they should be replaced with new ones. Do not allow the brackets to work loose as this is a very damaging condition and should never be allowed to exist. When rebabbitting the rear axle bearings be sure that they are square with the engine, otherwise it will not be propelled in a straight course.

If the axles are given reasonable care and management they will give the engineer less trouble than most other parts of the engine.

Front Axles.—Front axles are made in so many different styles and types that a description of them is almost out of the question. The support of these axles must be very strong and substantial and must be kept in shape or it will not be able to withstand the enormous stresses to which it is subjected. Not only are the wheels continually jerking and pushing this support but the steering gear puts an additional strain on it. The side stresses are also very severe, especially when the engine is turning a sharp corner. The bracket or support is likely to tear itself loose from the boiler should the front axles work themselves loose; therefore make it a practice often to examine the bolts holding the bracket to see that they are kept tight. In order that the engine will steer nicely do not allow any end play on the wheels and also see that they are kept the same distance from the center of the axle. Be sure that the wheels are well lubricated on the axles so as to prevent them from cutting. Another important item is to see that the brace used on the axle to keep it from tilting is in its proper place and well fastened. The steering chains should be so fastened to the axle that each chain will be the same distance from the center.

If the front axle support is attached to the smoke box by means of tap screws, these latter will finally become weak and pull out of the sheet, due to the corroding effect of the soot and ashes in the smoke box on the sheet and threads of the bolt. Whenever the axle pulls the bracket off the smoke box, about the best and safest way to repair it is to get a sheet of old boiler plate several inches larger than the face of the bracket, bore holes in it to conform with the holes in the bracket and form this sheet to fit the smoke box. Place the bracket back in position with the new sheet between it and the shell and fasten it with the proper size of bolts. Always place a large heavy washer on the bolts inside the smoke box to prevent them from pulling through the weak sheet. The washers can be of the bridge bolt type or made from a strip of heavy flat iron two inches wide and of a convenient length.

Steering Attachment.—One of the most widely used and simplest forms of steering gears is illustrated in Fig. 64.

CARE AND MANAGEMENT OF ENGINES

In this form of gear a roller to which chains are attached is placed in brackets which hold it at right angles with the boiler. These chains are attached to the roller in such a way that while one winds up the other is unwinding. They are also fastened one to each end of the front axle. It is obvious, therefore, that turning the roller will change the respective lengths of each of these chains which in turn

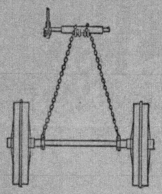

FIG. 64.—WORM STEERING GEAR.

will change the position of the front axle. In order that the operator may have control over this roller there is fitted to one end of the shaft carrying it a worm wheel. In mesh with the latter is a worm mounted on a shaft running to the platform and having a hand wheel fastened to its upper end. Turning this wheel will cause the roller to revolve slowly, which in turn will change the length of the chains, thus turning the axle. The chains are put on in such a manner that turning the hand wheel to the right will cause the engine to steer to the right and vice versa.

STEAM TRACTION ENGINEERING

They should be fitted with heavy springs to prevent severe shocks to the chains, roller and brackets holding them. They should be just tight enough so that one turn of the steering wheel will take up all the slack. If they should

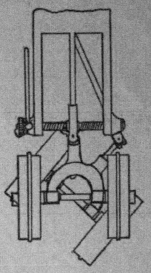

FIG. 65.—SCREW STEERING GEAR.

be too tight the engine will steer hard and if too loose it will be hard to keep the engine in the road, besides putting heavy strains on the whole steering gear. The roller shaft should not be allowed to get loose in the brackets nor the steering shaft to have much play in its bearings as this allows the worm and worm wheel to mesh improperly, in-

CARE AND MANAGEMENT OF ENGINES

ducing undue wear and allowing them a much better chance to strip the teeth on them.

Another style of steering gear known as the screw shaft device is illustrated in Fig. 65. In the construction of this style of gear the chains are replaced by a long heavy screw shaft. The turning of the hand wheel revolves this screw through the medium of the gears on the end of the shaft. A screw box is mounted on the screw shaft and connected to the front wheels by a large heavy central arm. The turning of the shaft shifts the box to one side or the other, thus turning the front axle the desired amount required to steer the engine. The slack and jerking so common with the chain form of steering gear is entirely obviated in this gear.

Water Tanks.—All traction engines are, or should be, equipped with some kind of tank to carry a supply of water while the engine is moving. The location of the tank varies according to the make of engine but the most usual places are either on the front end of the boiler, on the side of the platform, along the side of the boiler, or underneath the platform. The tank or tanks are almost universally fitted with a tight cover to keep out the dirt, but no matter where located or how covered they will accumulate dirt to a certain extent and therefore should be periodically cleaned. They are usually fitted with either a hand hole or plug placed near the bottom of the tank, through which the dirt can be removed by scraping and washing. If they are cleaned regularly the dirt can usually be washed out with very little trouble. Some makers arrange to have the tank filled by means of a jet while others use a tank pump or buckets to do the job. If injectors are used to feed the boiler and the tank is filled by means of a jet, care must be taken that the water does not get too hot for the injector to handle satisfactorily. This also applies to a tank which is located on the top of the boiler.

STEAM TRACTION ENGINEERING

The old style wooden wagon tank for hauling water to the engine is fast going out of use. The tank should be mounted on a good strong wagon fitted with a good tank pump, suction hose, discharge hose and nozzle and kept clean by frequent washings. An old tank is about the most aggravating thing to have around an engine. It keeps the axles on the wagon wet, which soon rusts them and rots the wood and causes a breakdown when the water is needed the most. The water hauler should not only clean the tank frequently but should keep a strainer at all times on the end of the suction hose. An old baking can, large enough to slip over the hose and punched full of holes will provide a good substitute for a strainer. If a coarse strainer is used on the suction hose and the water is full of fine stuff which passes through it, take an ordinary bran sack without any holes in it and tie the mouth of it around the pump spout. This will strain the water nicely and prevent pieces of moss or straw entering the tank.

Tender.—A tender is a necessity to the well-equipped engine, especially with large engines engaged in heavy traction work or where considerable moving is being done. It is made in numerous styles, having the same general principles according to the design of the engine to which it is attached. The latest and most popular style of tender is an enlarged platform carrying a water tank and a place for fuel. This is a handy and compact form of tender. The engine will be shorter than would be the case with the wheeled tender, besides not requiring an extra truck to support it, and there will be less trouble in steering the engine. It is open to several minor objections; for instance, on small engines it puts a very heavy strain on the rear part of the engine as the drive wheels are compelled to carry all of this increased load. Also it is very close to the ground in some makes of small engines and the platform is usually much crowded.

CARE AND MANAGEMENT OF ENGINES

Another style of tender widely used is what might be called the "two-wheeled tender." This type is mounted on a two-wheeled truck which is attached to the engine by means of a draw bar. Chains or cables run from the axle of the tender to the front axle of the engine in such a way

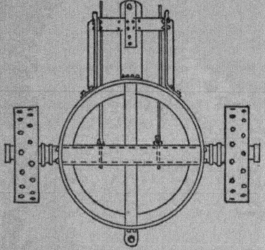

FIG. 66.—TWO-WHEELED TENDER TRUCK.

that when the front wheels are turned to the right, the tender wheels are turned to the left, thus making the tender track directly behind the engine, whether going ahead or backing up. Side links are always attached so that the tender cannot be dragged out of line with the engine by a side pull. A draw bar is fastened to the rear of the tender for coupling on loads. The cables or chains guiding the tender wheels must not be allowed to get very loose; or the tender will not track properly, while on the other hand if they are too tight they will cause the engine to steer hard.

The water tank is usually mounted directly on top of the truck with the coal box on top of the tank. The truck of one of these tenders is shown in Fig. 66. The wheels and steering gear of a two-wheeled truck should be well oiled.

Storage of an Engine.—Every engine should be given the same attention before being placed in storage after a season's run as when it was at work if it is desired that it give good service and last a long time. The practice of running a traction engine alongside of a barn or shed when through threshing, letting it cool down and then removing the water, as so many do, is harder on the engine than to run it continuously.

Directions as to what was to be done with the boiler so that it would be in good shape the next season were given on page 66 and therefore will not be repeated. It would be a good plan to read this section again if for any reason the operator has forgotten what he was supposed to do.

The following directions refer to the preparation of the engine mechanism for storage: First drain pipes thoroughly of water and clean them of all deposits. Remove the packing from all stuffing boxes and oil the valve rods, spindles and piston rods with a coat of cylinder oil. The cylinder head should be removed and the cylinder walls given a good coat of cylinder oil; also remove the steam chest cover and give the valve and seat the same kind of treatment. Clean the grease from the engine and boxes and if the engine stands out of doors, it is a good plan to tie a piece of sacking or canvas over the boxes to keep out the dirt and sand. The engine and boiler should be repaired if necessary at this time so that the machine will be ready to start again at a moment's notice. Every engine should have a shed to stand in, just the same as a horse.

CARE AND MANAGEMENT OF ENGINES

Handling a Traction Engine.—*On the Road.*—There are five things which the engineer will have to learn in order that he may operate a traction engine on the road successfully. He may be entirely familiar with steam and steam engines and may be a very good engineer in a stationary power plant and yet the very first time he runs a traction engine he may get himself and the engine into difficulties.

1. The engineer must have a thorough knowledge of how to handle the throttle properly. There is a particular knack in the proper handling of this valve which will differentiate the poor operator from the experienced traction engineer. The experienced man will turn his engine around and couple up the load in a short space of time without either making a false move or jerking his engine and load all to pieces in doing it. The other type of man will do everything with a rush, couple up with a jam, smash someone's fingers in doing it and always leave with a jerk straining the engine and the load that it is pulling and probably breaking a coupling or a gear. A beginner will do well to go slow for a while until thoroughly acquainted with the peculiarities of the particular engine which he is operating. In backing and coupling, open the throttle only a little so that the engine is under perfect control. In starting an engine either forward or backward open the throttle slowly until the engine has obtained the desired power which it requires to do the work. The amount of this opening can only be found out in the school of experience.

2. The steering wheel usually gives beginners more trouble than anything else when the engine is moving on the road, as they have a practice of turning the hand wheel entirely too much. It must be remembered that one turn too much in one direction necessitates two turns in the opposite direction. Always keep your eyes on the front wheels of the engine and on the road ahead. When making

STEAM TRACTION ENGINEERING

a difficult or bad turn it is much the best plan to go slowly and carefully, as then the engine can be stopped easily, thereby preventing it from getting the operator into serious trouble. A traction engine being a heavy piece of machinery will sooner or later land in a mud hole, sand pit or other soft place in the road. In order to avoid just such places, the engineer must keep a sharp lookout on the road while going. To most engineers, especially beginners, the engine seems to have a faculty for hunting out all the soft places in the road.

3. How to get out of a hole is sometimes a very vexing question and a lot of patience is spent and time consumed in the attempt. If one wheel has a good solid footing do not spoil it by trying to pull the engine out, but instead give the other wheel something to adhere to. One of the best things for this purpose is a heavy chain wrapped around the wheel which is stuck, in such a manner that it will not slip. Failing to obtain a chain, pieces of old fence posts, rails, planks or even bushes are good things to put under the wheels. If unable to pull out at the first attempt put more stuff under the wheels and try again. If the wheels of the engine have sunk down in the soft earth until the fire box rests on the ground or the wheels refuse to move at all, the only thing to be done is to dig out the engine.

When pulling over either wet or dry sand go slowly and steadily so as not to break the footing of the engine. Do not rush at a sand bed nor get the idea that enough speed can be developed in a traction engine to push it over a sand bed, for the amount of speed that can be developed is not enough to get the engine very far in sand. Should the wheels break through and go to slipping, stop the engine instantly so that the wheels will not get buried too far; then get something under them, such as either straw, hay, brush or boards, in order to give them a better foot-

CARE AND MANAGEMENT OF ENGINES

ing. Patience and good judgment will get an engine out of a hole or sand bed quicker than all bucking and tearing.

4. Whenever the engineer comes to a hill and he has any misgivings about the clutch slipping he should use the clutch pin instead. With a good heavy fire and a fair head of water a traction engine should be able to go up any hill of fair size without stopping. Do not attempt to rush up a hill, especially if the engine is loaded and pulling hard, as it may cause the boiler to prime, which will necessitate stopping the engine. See to it that the safety valve does not blow when making a hard pull as this will result in the boiler priming.

5. Last, but not by any means unimportant, watch out for bad bridges and culverts. Many a man has either lost his life or been crippled in a bridge wreck. Never pull on to a strange bridge or culvert without making a thorough examination of it. If there is any doubt about its holding the engine go around it, but if this cannot be done then thoroughly brace it before attempting to cross. Avoid all sudden shock and quick stops when crossing a bridge.

Do not get excited or rattled as this is one of the worst traits which an engineer can have unless it is to get drunk on duty; and remember that a man who is intoxicated is not fit to handle an engine anywhere.

Placing the Engine in Line with the Machine.—In order to become a successful operating engineer of a traction engine, it will be necessary to be able to place the engine in line with the machine to be driven quickly and easily. Some engineers will slam their engine back and forth four or five times and then not be in line with the machine by a yard. In lining up with a machine if the engine is run slowly enough the work can be done without much trouble. The school of experience is the only source from which the engineer can draw in order to learn the secret of correct and easy lining up of an engine and machine.

If there is no wind to throw the belt out of line, or if it is in the same direction as the belt, the engine should be set in a direct line with the machine, i.e., a line drawn touching both the edges of the flywheel should strike the edge of the pulley of the machine on the same side, and one drawn past the edge of the pulley should strike the edge of the flywheel. This will allow the belt to run true and in the center of both pulleys. When a side wind is encountered it may be necessary to set the engine out of direct line, according to the pressure of the wind and how long a belt is being used. With a heavy wind it is sometimes necessary to set the engine as much as three feet out of line. The engine must be set over to windward or against the way the wind is blowing. The amount of offset will all depend upon conditions and the engineer must learn it from his own experience.

Stationary Work.—By this is meant any class of belt work which is done with a traction engine. This subject of the proper handling of an engine has already been covered pretty thoroughly in other parts of this book but it would not be out of place to recapitulate them here:

1. Keep the engine in good shape all the time and keep it clean.

2. Hold an even pressure of steam in the boiler as long as possible and try to keep the water level constant while running.

3. Where possible hook the reverse lever up as close to the center as is possible to do the work required of the engine.

4. If doing light work with the engine do not let the steam drop but keep it up to the full pressure at all times as it saves both fuel and water.

CARE AND MANAGEMENT OF ENGINES

5. Keep the engine blocked solidly as vibration is one of the worst enemies of smooth running machinery.

Above all else, attend strictly to the business of watching the engine and always be near it so that if anything happens it can be quickly stopped.

Plowing and Heavy Hauling.—Plowing and heavy hauling are the heaviest and hardest strains that can be put on a traction engine and therefore require close watching and the engineer's best judgment.

1. Do not carry an excessive amount of water and keep the boiler clean by washing it out as often as necessary.

2. Keep the steam pressure even and always carry the pressure up to the point recommended for good work.

3. Keep a steady even speed and do not try to rush things. To keep moving all the time is what counts in the end.

4. Keep the gearing well lubricated and in line, the boxes in good shape and the steering gear in perfect order.

Plowing especially is a heavy constant load on the engine and a sharp lookout should be kept for loose bolts and boxes, hot bearings or anything which is liable to give trouble.

CHAPTER VI

VALVE GEARS AND VALVE SETTING

In order to understand this chapter, what was said under the heading of Eccentric on page 117 should be read over. The purpose of this chapter is to give the engineer an understanding as to valve gears and how properly to set them. A knowledge of the different types of valves and how they should be set is of vital importance to every successful engineer.

As before mentioned the eccentric, or something which takes its place, is used to drive the valve, therefore its location on the shaft must be exact and its throw properly proportioned for the travel of the valve. The throw is determined by the designer of the engine; therefore the engineer will not be called upon to determine it but will have to see that the eccentric is properly located.

In the following short treatment of the peculiarities of the valve, the plain D slide valve will be described, as what applies to it will also apply to all classes of valves as to movements and functions. A sectional view of this type of valve on its valve seat is illustrated in Fig. 67, in which AA are the steam ports; BB the bridges; C the exhaust cavity; DD the valve seat; E, F, G, H the valve faces; II the outside lap; JJ the inside lap; MN is the center line of the valve and valve seat on port face. The valve is shown in its central position, i.e., the center line N of the valve coincides with the center line M of the valve seat.

Lap.—The outside and inside lap is shown in the above Figure (67) as II and JJ respectively. The lap is the

VALVE GEARS AND VALVE SETTING

amount which the valve laps over the port openings when such valve stands in its central position. The lap on the valve is necessary to provide means for cutting off the steam before the piston reaches the end of its stroke and also to close the exhaust port in the same manner. The outside lap is given the valve to cause it to stop the admission of steam through the ports earlier in the stroke, thus allowing the steam to push the piston the remainder of the

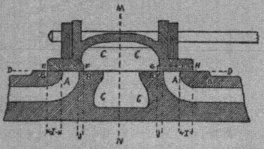

FIG. 67.—PLAIN D VALVE.

stroke by expansion. If no outside lap were given the steam would enter the cylinder the full length of the stroke. The inside lap is given to the valve to allow it to close the exhaust port a little before the piston reaches the end of its stroke. This causes the piston to compress the steam remaining in the cylinder and form a cushion partly to relieve the engine from the jar of suddenly bringing the piston to a stop, previous to starting on its return to the other end of the cylinder. This point of exhaust cut-off is called the *point of compression*.

Lead.—Lead is the space or opening between the outside edge of the valve and edge of the port on the steam end of the cylinder when the engine is on its dead center. It is allowed in order that a small amount of live steam may enter the cylinder just before the piston reaches the end of its stroke. This helps the compression to form a cushion which enables the engine to pass the center without a jar.

Too much lead is a source of weakness in an engine as it will allow the steam to enter the cylinder quite a bit before the piston reaches the end of its stroke, thereby forming a back pressure in the cylinder which tends to prevent the engine from passing its center. This will cause the engine to bump and pound, making it hard on the engine besides materially lessening its power. Too little lead will not allow sufficient steam to enter the cylinder ahead of the piston to start the reciprocating parts promptly on their return stroke and to form enough of a cushion to stop the inertia of the rapidly moving parts, with the result that the engine will pound the crank pin and crosshead brasses, thus making it difficult to keep the engine running smoothly.

After reading the above the engineer will naturally come to the conclusion that it is very important to have the lead right about an engine and he will be anxious to know when the lead is properly set and if not how he may make it right. If the engine had the valve properly set in the first place, as in the case of a new engine which has just left the factory, it will be right as long as the adjustment is not altered, but wear will gradually change it, or one or another part may slip, requiring the resetting of the valve. As different makes, different styles of engines, valves and valve gears vary so much, it is next to impossible to give any definite rule for the exact amount of lead required. For instance, one engine may have ports eight inches long and a half inch wide while another may have them four inches long by one inch wide. It is evident that both sizes of port have the same area and will supply the same size of

VALVE GEARS AND VALVE SETTING

cylinder, but the short, wide ports will require a longer valve travel to open them than the one having the long and narrow ports. From the above it will be seen that without knowing the area of the piston, size of the port, speed, valve travel, etc., it will be a very hard matter to give the correct amount of lead.

Engines which have a reversing valve gear using only one eccentric have the lead fixed at the factory, the amount of which cannot be changed. If the gear is set for more than the given lead on one motion, it will not have sufficient lead when running on the other motion. This rule applies to all engines that have a single eccentric reversing valve gear. On engines using the link reverse or those not equipped with any reverse gear but a plain eccentric this rule does not apply as the lead is changeable by moving the eccentric.

Expansion.—Steam is an elastic gas, and has a great tendency to expand. Steam at atmospheric pressure occupies seventeen hundred times the space that the water occupied from which it was generated. When confined under pressure, as in the cylinder or the boiler, it always tends to expand itself to the fullest extent. The confining of the steam and the increased temperature above the boiling point determine the pressure of the steam. Owing to this expansive force of the steam, a great amount of fuel and water can be saved by using the steam in this manner, i.e., by cutting off the steam during the early part of the stroke and allowing the steam's expansive force to complete the stroke of the piston. The most common points of cut-off are one-quarter, one-half, and three-quarters of the stroke of the piston. In a few cases three-eighths and five-eighths of the stroke are used. To illustrate what this expansive force of the steam amounts to in an engine, suppose there are two engines having the same size of cylinder, the first admitting steam during the full length of the stroke and

STEAM TRACTION ENGINEERING

the second allowing the steam to be cut off after the piston has traveled half the length of the stroke. Suppose that the steam pressure in both the engines is 100 pounds and that the area of the piston is 50 square inches. The first engine will, therefore, have a total pressure in the cylinder of $50 \times 100 = 5,000$ pounds, while the other engine which cut off the steam at half the stroke will only require half as much steam and the average pressure throughout the stroke will be 84.75 (take from the Table of Average Pressures, given below) pounds per square inch. This engine therefore has a total pressure of $50 \times 84.75 = 4,237.5$ pounds. This gives only about one-seventh less power by using one-half less weight of steam. Below is given a table of average pressures of steam on the piston per square inch when steam is cut off at one-fourth, one-half and three-fourths of the stroke, commencing with a boiler pressure of 70 pounds and advancing by 5 pounds up to and including 140 pounds.

TABLE OF AVERAGE PRESSURES

Boiler Pressure	70	75	80	85	90	95	100	105
Cut off at ¼	41.5	44.5	47.5	51	53.5	56.5	60	62.5
Cut off at ½	59.25	63.5	68	72	76.25	80.5	84.5	87
Cut off at ¾	67.5	72.25	77.25	82	87	91.25	96.5	101

Boiler Pressure	110	115	120	125	130	135	140
Cut off at ¼	65.5	68.5	72	75	77.5	81	83.5
Cut off at ½	93	97.25	101.5	105	110	114.5	119
Cut off at ¾	106	110.5	116	119.5	126	131	136.5

This table is useful in figuring the horsepower of an engine (page 282) but in so doing the back pressure must be deducted in order to obtain the mean effective pressure.

When a very small exhaust nozzle is used a higher rate would have to be deducted.

Direct and Indirect Valves.—A valve is said to be direct when it opens the left steam port by moving to the right and closes it by moving to the left. An indirect valve is

VALVE GEARS AND VALVE SETTING

one which operates in the reverse to the above, i.e., opens the left port when it moves to the left and closes it by moving to the right. The plain D slide valve already described is a typical example of a direct type of valve, while the double-ported piston valve illustrated in Fig. 68 is the most common form of indirect valve.

This valve consists of a hollow cylinder sliding inside of a cylindrical valve seat and having the ports PP extending

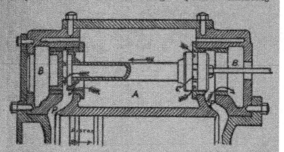

FIG. 68.—DOUBLE-PORTED PISTON VALVE.

entirely around the valve. The steam is admitted into the central chamber A and the exhaust escapes at the two ends BB. As shown in the figure the piston is about to start to the right, the valve is moving to the left, opening the left steam port, and allowing live steam to enter the port past the inside end of the valve. To give a quicker port opening by a larger admission of steam the latter is allowed to pass into the center of the valve by means of the channel C and thence into the left port. The exhaust escapes through the right steam port into the chamber B. It is plain that the direction of motion of an indirect valve is precisely opposite that of a direct valve and therefore the eccentric must be set exactly opposite to the position it

STEAM TRACTION ENGINEERING

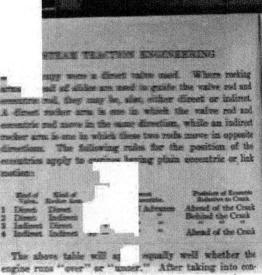

any were a direct valve used. Where rocker arms or shaft of slides are used to guide the valve rod and eccentric rod, they may be, also, either direct or indirect. A direct rocker arm is one in which the valve rod and eccentric rod move in the same direction, while an indirect rocker arm is one in which these two rods move in opposite directions. The following rules for the position of the eccentric apply to engines having plain eccentric or link motion:

Kind of Valve	Kind of Rocker Arm		Position of Eccentric Relative to Crank
1 Direct	Direct		Ahead of the Crank
2 Direct	Indirect		Behind the Crank
3 Indirect	Direct		" " "
4 Indirect	Indirect		Ahead of the Crank

The above table will apply equally well whether the engine runs "over" or "under." After taking into consideration which direction the engine is to run, set the eccentric so that it reaches a given point either ahead or behind the crank as given in the table. The *angle of advance* is the amount the eccentric must be set ahead or behind the ninety-degree position with the crank, this amount being equal to the lap and lead of the valve.

The Plain Piston Valve.—This valve, which is a direct one (Fig. 69), differs from the double-ported one which was shown in the previous illustration in that the steam is only admitted past the outside edge and is not allowed to pass through the valve. In all other respects the two valves are identical. The lead may be easily seen and regulated in a double-ported piston valve by removing the steam chest cover over A, while to adjust the lead on a plain piston valve the plates on the *end* of the steam chest must be removed. In the former type of valve the lead is adjusted from the *inside edge of the valve* while the adjustment of the latter is from the *outside edge of the valve*.

VALVE GEARS AND VALVE SETTING

The plain piston valve has relief valves held by springs over each steam port to prevent breakage should a considerable amount of water enter the cylinder. Both types of piston valves are nearly balanced because the steam entirely surrounds them and therefore they require very little power to drive them. The plain D valve is not so balanced.

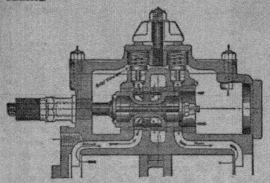

FIG. 69.—PLAIN PISTON VALVE.

The Allen Valve.—This valve is a modification of the plain D valve, being double ported. The double-ported valve is used in engines which are to be operated at a high speed or where it is desirable that there be a quicker opening and closing of the ports than can be obtained with the ordinary valve. This valve illustrated in Fig. 70 has a passage A cast in the valve. The shoulders BB of the valve seat are so constructed that when the edge m of the valve is just even with edge J of the port, the outer edge p of the passage A will be just even with edge n of the shoulder B at the outer edge of the valve seat. When the

STEAM TRACTION ENGINEERING

valve moves a little to the right into the position shown in the cut, steam will enter the port directly between the edges J and n as in the case of the ordinary valve, and at the same time the edge p of the passage A has moved past the edge n of the valve seat, thus allowing steam entering the passage A to enter directly to the left steam port. The exhaust is controlled in the same manner as with the ordinary valve. In adjusting the lead the outside edge only is considered and is usually made about one-half as much as with the single-ported valve. The steam ports of this valve

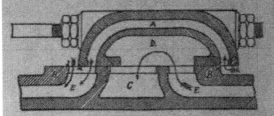

FIG. 70.—ALLEN VALVE.

have a slightly wider opening than with the ordinary type of valve.

The Double-ported Balanced Valve.—The distinctive feature about this valve as illustrated in Fig. 71 is that it has two openings, C and D, to each steam port. The valve also has two passages, B and B, which extend through it and communicate with the steam chest A. In the position shown in the cut the valve is opening the left steam port. The steam enters the passage C past the edge of the valve and also enters the passage D through the opening in the chamber B, meanwhile the exhaust is escaping from the right port into the chamber E and exhaust port G at the same time it is escaping through the passage C into the

VALVE GEARS AND VALVE SETTING

exhaust port. The plate F is called the balance plate and is held in position by the studs in the steam chest cover. This plate is mechanically supported and prevents the direct action of the steam in the steam chest A from pressing on the back of the valve, and this enables it to move with less effort than without its use.

It is evident that this valve will give double the port opening with the same valve travel as that of a plain D slide valve. The lead on this valve is set in the same way as

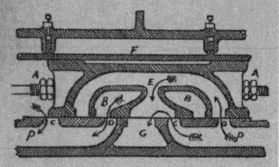

FIG. 71.—DOUBLE-PORTED BALANCED VALVE.

that of the Allen or plain D valve but it requires one-half as much lead as the D valve, or about the same amount that is required for the Allen valve.

The Giddings Valve.—This valve shown in Fig. 72 is both a double-ported and a balanced one. By referring to the cut it will be noticed that the valve seat contains five ports. Live steam enters the central port from the steam pipe and goes from thence to the central valve chamber and ports BBB. There are two exhaust ports, one on each side of the live steam port, and the steam ports to the cylinder are the outside ports FF. In the illustration the valve is shown in position as the piston just starts to travel to the right, the left steam port being open slightly more than the amount of the lead. Steam enters from the central steam port through the cavity and ports marked B and also through the passage A to the steam port F. The ports in the valve are arranged so that the inside edge of

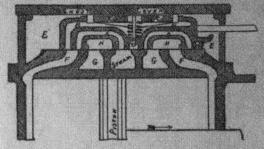

FIG. 72.—GIDDINGS VALVE.

the port A is in line with the edge of the central steam port at the same instant that the inside edge of the passage B comes in line with the outside edge of the steam port F, thus giving a double-ported opening. The exhaust is passing from the right steam port F into the right exhaust port G through the passage H in the under side of the valve.

This double-ported feature gives double the port opening for the same valve travel as that of the ordinary D slide valve. It also gives a much larger wearing surface to the valve and seat, making it very durable. The balancing feature is accomplished by having a circular portion cut out of the back of the valve large enough to take up

VALVE GEARS AND VALVE SETTING

cover EE, which fits over the valve, has the inside highly finished and ground and when in place completely covers the valve. The snap rings BB make a steam-tight joint at each end of the cover and valve, so that steam cannot get back of the valve, thus relieving it of the pressure of the steam on its back and making it a balanced valve. Very little power is required to move the valve since all the friction present is that caused by the weight of the valve

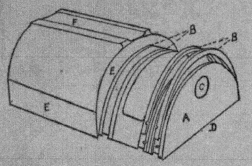

FIG. 75.—GOULD BALANCED VALVE.

and the slight pressure of the snap rings. Springs are used between the steam and valve chest covers, to prevent an accident in case water enters the cylinder. They hold the valve cover in place, but allow any other pressure to raise the valve and cover, and when such pressure is relieved the springs instantly return the valve and cover to their proper place.

In setting these valves the cover should be lifted off, if the ports cannot otherwise be seen, and the valve set as regards lead in the same manner as the D slide valve. In replacing the valve cover be careful that the snap rings

STEAM TRACTION ENGINEERING

are in their proper place, and when replacing the lid of the steam chest see to it that the springs are also in their proper place between the steam chest and the valve cover.

The Newton Balanced Valve.—This is another balanced valve which operates on the same principle as the D slide valve. Two views of this valve are illustrated in Fig. 76. The left-hand view (A) shows the back of the valve with the valve cover removed. In this view the body of the valve A is semicircular in shape, the flat part being the valve face, while the circular back is fitted with grooves in which snap rings B are fitted. A cover fits snugly over

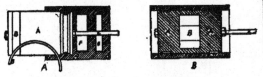

Fig. 76.—Newton Balanced Valve.

the valve and the rings form a steam-tight joint between the cover and the back of the valve. The valve is balanced since the pressure of the steam is prevented from pressing on the back of the valve and therefore it requires very little power to drive it, the friction having been nearly all removed. In the illustration, A, C is the valve stem; DD the duplicate port face; E the steam port, and F the exhaust port. In the second view (B) shown A is the valve face; B the exhaust cavity; C the valve rod, and DD the circular valve cover.

These valves are made especially to replace the plain D slide valve in the field, as no special tools are required to install it in any engine. The use of a duplicate port face obviates the necessity of taking an engine to a machine shop to install the valve. A special packing is placed be-

VALVE GEARS AND VALVE SETTING

tween the old port face and the new one. The valve cover fits over the valve and on to the edge of the duplicate port face, the whole being held in place by the steam chest cover. The Newton is a direct valve and is set in exactly the same manner as the plain D slide valve.

The Non-leak Balanced Valve.—As will be seen from a study of Fig. 77, this valve is also an improved form of slide valve, being derived closely from the plain D valve. This valve, like the Newton, is also made to be easily attached in the field without going to the expense of having

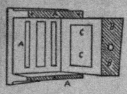 

FIG. 77.—NON-LEAK BALANCED VALVE.

the valve fitted to the old port face at a machine shop. In the illustration A is the detachable or duplicate port face, B the valve, C the exhaust cavity, and D the balance plate.

It will be seen that the valve is made wedge shaped with the narrow side down. The balance plate is carefully set against shoulders in the duplicate port face and fits snugly the valve back, both plate and back being scraped to a steam-tight fit. No rings or springs are used in the construction of this valve and therefore there is no friction except that due to the weight of the valve. It can be clearly seen that the wedge shape of the valve will automatically take up all wear and keep the valve tight and perfectly balanced at all times. There is a small space between the balance plate and the removable back of the port face and frame, which allows the balance plate and valve to rise from the port face in case of water in the cylinder. The valve will be instantly returned to its seat as soon as the excess pressure is removed by means of the pressure of the live steam on the back of the balance plate. The valve and port face are held in position in the steam chest by the steam chest cover. The valve is a direct valve

FIG. 78.—BAKER BALANCED VALVE.

and is set and adjusted in exactly the same manner as the plain D slide valve.

The Baker Balanced Valve.—This valve with its duplicate port face is illustrated in Fig. 78. The valve as shown in the cut is almost entirely withdrawn from the port face. It is a plain piston valve of the direct type so arranged as to be easily attached to an engine in the field, the port face being designed to fit over the old port face without any careful planning. No description of the valve itself is necessary as the description of the plain piston valve will cover all the points about this valve. The cage for the valve, which constitutes the new valve seat, is the same casting as the duplicate port face. The valve and valve

VALVE GEARS AND VALVE SETTING

cage, which includes the duplicate port face, are held in place by the steam chest covers. Since this valve is cylindrical in form and the steam acts with equal pressure on all sides it is in perfect balance. Steam is admitted past the outside edges and exhausts into the central portion of the valve. Therefore it is a direct valve and will easily take the place of the plain D valve, the same valve travel being used as the old valve and driven by the old valve gear without the necessity of any change. Setting the valve and adjusting the lead is accomplished in the same manner as with the direct piston or plain D valve.

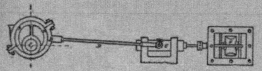

FIG. 79.—PLAIN ECCENTRIC VALVE GEAR.

Valve Gears.—There are a number of valve gears on the market and nearly all of them are used on the different makes of farm engines. The simplest form of valve gear is illustrated in Fig. 79 connected to a plain slide valve. This valve gear is not reversible and is, therefore, only used on portable engines as they only run in one direction. The gear can be reversed but only by resetting it. In the cut A is the eccentric; B the eccentric yoke; C the main shaft; D the eccentric rod; E the valve slide, and F the valve.

The horizontal line indicates the dead line of the engine and the vertical angle the ninety-degree angle with crank pin on dead center. This line is the ninety-degree position of the eccentric in relation to the crank, but it does not indicate the correct position of the eccentric, as the true position is either ahead or behind this line, according to the amount of the angle of advance. The *angle of advance* may be described as follows: If the valve had no outside lap the eccentric would be set exactly ninety degrees from the crank center line. In this position the valve would be just in line with the edge of the steam port so that the slightest movement of the valve would admit steam to the port. But as all valves of the present day have outside lap, then if the eccentric were in this position the ports would not be opened until the crank pin had gone measurably past the center. Now, on account of both outside lap and lead, the valve must be moved from the central position in the direction in which the engine is to turn, the amount of the lap plus the amount of the lead. To accomplish this the eccentric is turned on the shaft sufficiently to move the valve the proper amount. This places the correct position of the eccentric either ahead or behind the ninety-degree position according to the kind of valve and rocker arm used. The amount of this distance is known as the angle of advance.

Directions for Setting the Plain Eccentric and a D Valve.—In order to set a valve it is first necessary to center the engine, which requires that the inside and outside dead centers of an engine be determined. An engine is said to be on its dead center when the piston rod, crosshead pin, connecting rod and crank pin are in a true straight line. There are two dead centers of an engine, one when the crank pin is on the center nearest the cylinder, which is known as the head or inside dead center, and the other when on the center farthest from the cylinder, known as the crank or outside dead center.

The dead center cannot be determined accurately enough by eye to set the valve properly.

To center an engine accurately, provide a tram eighteen inches long, made of a quarter-inch steel rod both ends of which are bent at right angles and both ends sharpened to

VALVE GEARS AND VALVE SETTING

a point. Turn the engine until the crosshead is nearly at the end of its travel and make a scratch mark on the guides exactly at the end of the crosshead shoes. If it is hard to make a mark that is accurate, in this manner, with a rule or a straight edge make a mark across both the guides and crosshead. Place one end of the tram, say the left-hand end, on some convenient stationary part of the engine frame, so that the other or right-hand end will reach the face of the flywheel (Fig. 80). Make a prick punch mark at the point where the left end of the tram is placed and with that end of the tram in this punch mark make a scratch on the face of the flywheel with the other end of the tram.

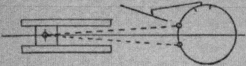

FIG. 80.—CENTERING AN ENGINE.

Now turn the engine past the center until the marks on the crosshead and on the guides are exactly in line again with the crank pin on the other side of the center. With the engine in this position place the left end of the tram again in the punch mark, and make another scratch on the face of the flywheel with the other end. Carefully measure the distance between these two scratch marks on the face of the flywheel and make a plain prick punch mark exactly midway between them. When the engine is turned so that the outer or right end of the tram fits this mark, the inner or left end being in the punch mark on the engine frame, it will be on the true dead center. Both centers are found in exactly the same way. With a disk crank engine just as good results can be obtained by having the lines and punch marks placed on the disk instead of the flywheel. The engineer should take his time in finding the true dead centers as the accuracy of the valve setting depends largely upon the accuracy of the centers.

Before starting to set the valve have all the lost motion taken up as much as possible in the connecting rod boxes, crosshead, main boxes, eccentric yoke and all the valve connections, and have the valve rods of the correct length. In setting the valve turn the engine over on its head dead center, remove the steam chest cover and loosen the eccentric. If the eccentric and shaft are marked for the correct position of the eccentric all that needs to be done is to set the eccentric to these marks and adjust the lead, but should there be several marks or none at all the following plan should be used. Carefully measure the valve and seat and then set the valve in its central position, i.e., so that it covers both steam ports equally. Make a light scratch mark on the port face, so that the valve can be placed back in this position again easily should it become moved. If a rocker arm is used on the valve stem, place it in a vertical position, or if a slide is used put it in the center of its travel. Now then adjust the length of the valve stem so that it is the correct length to connect the valve with the rocker arm or slide when both are in the above-mentioned central positions. Next turn the eccentric on the shaft with the point of its greatest throw either up or down according to the way the engine is to run, and the kind of valve and rocker arm used. The table giving the position of eccentrics will be found on page 180.

Suppose the engine is to run "over," i.e., the crank pin will travel *away* from the cylinder when on its top position, and a direct rocker arm and valve is used. In this case turn the throw of the eccentric exactly straight up, in which position the eccentric is at ninety degrees or "quartered." The valve and rocker arm being in their central position, adjust the length of the eccentric rod so that it

VALVE GEARS AND VALVE SETTING

will be of just the correct length to connect the rocker arm and the eccentric when all are in their central positions. With the engine still on the "head" dead center, turn the eccentric *ahead* enough to open the head-end steam port the required lead. The amount of the lead will vary from one-sixty-fourth to one-eighth of an inch according to the size of the cylinder and the speed of the engine. The larger the engine or the higher the speed the greater the amount of lead used. Good results are obtained for ordinary sized engines running at a moderate speed, if the lead is three-sixty-fourths of an inch. This amount can be gaged with the back of an ordinary knife. When the required amount of lead is obtained fasten the eccentric and turn the engine over until it is on its other center. Should the lead be not the same at this end adjust it by moving the eccentric the proper way to correct half the error, and then turn the engine back to the other center in order to check the amount of lead in this position. When the proper lead has been obtained at both ends fasten the eccentric securely in that position.

The valve should be examined to see that the fastenings holding it in position on the valve stem are tight and that the valve is also free to raise from its seat without binding on the valve stem or nuts holding it. It should not have over one-thirty-second of an inch end play on the stem, as this amount will give plenty of room to allow the valve to rise from its seat. The job will be completed upon the replacement of the steam chest cover.

Reversing Valve Gear.—Locomotive, marine, hoisting and traction engines are required to run in either direction over or under and for such engines it is necessary to provide some means by which they can be easily, quickly and accurately reversed. All traction engines and a few portable ones are equipped with some form of reversing valve gear. Some of these mechanisms are arranged so that

STEAM TRACTION ENGINEERING

the cut-off may be varied from one-quarter to nearly full stroke while others only reverse the engine and are arranged to cut off at one point in either direction that the engine may be running. The point of cut-off with such gears is usually about three-quarters of the stroke when the engine is running in either direction.

Reverse gears may be divided into three distinct groups:
1. Double eccentric—cut-off changeable.
2. Single eccentric—cut-off not changeable.

FIG. 81.—LINK REVERSE.

3. Gear and crank—cut-off changeable and not changeable.

Each of these reverse motions will be taken up in turn, a short explanation given of them and the method to be followed in setting them.

The Link Reverse.—This reverse is also commonly known as the double eccentric or the Stephenson valve gear. It derives the latter name from George Stephenson, who first applied it on locomotives in 1833. The link motion of today differs very little from the one first designed by Howe and used by Stephenson.

The outline of the link reverse is illustrated in Fig. 81. As can be plainly seen, there are two eccentrics and rods used which are connected by joints to a link. In the cut. C is the engine crank; D, the main shaft; A, the forward

VALVE GEARS AND VALVE SETTING

motion eccentric; B, the back motion eccentric; E, the forward motion eccentric rod; F, the back motion eccentric rod; L, the link, and V, the valve stem. The eccentric rods, E and F, are fastened to the link, L, by the pin bearings, d and f, and the link block is pivoted to the valve stem by the pin, a. This block fits in the slot of the link and when the link is raised or lowered the block will be changed from one end of the link slot to the other. A bracket is placed in the center of the link, to which the lifter rod, M, is pivoted by the pin, C, and this rod is also pivoted to the bell crank, N, which is attached to the reverse lever, P, by the reach rod, O. The reverse lever works in a quadrant, T, which is fitted with notches for holding the lever in the different positions.

It is plain that moving this lever from one end of the quadrant to the other will raise or depress the link. When the mechanism is in the position shown in the cut the eccentric rod, E, is in line with the valve stem and the action is precisely the same as an engine equipped with a single eccentric. The eccentric, A, will govern the steam distribution and B will have no effect whatever on the valve. Consequently the engine will run in the direction indicated by the arrow. If the reverse lever is moved to the other end of the quadrant, the lifting links, O, N, M, will raise the main line, E, until the eccentric rod, F, is in line with the valve stem and the eccentric, B, will then control the motion of the valve. Since the eccentric must be ahead of the crank in direct valves, it will move in the direction opposite the arrow.

Besides reversing the motion of the engine the link will also perform two other important duties, the first of which is that when the link is placed in mid gear, i.e., when the block, W, is exactly in the center of the link and the reverse lever in the center notch of the quadrant, the valve stem, V, is acted on equally by both eccentrics and the

STEAM TRACTION ENGINEERING

travel of the valve is so little that the engine will not run in either direction. The second duty that the link will perform is that when it is raised from its mid-position part of the way—the link block, W, about one-fourth of the distance between d and f—then the valve rod would be acted on mostly by the eccentric, A, but it will also be affected somewhat by the motion of the eccentric, B. The result will be that the valve rod will travel a shorter distance than it would if it was acted on by the eccentric, A, alone and will therefore cause the valve to cut off earlier. The compression in the cylinder will be increased by doing this as well as a saving in fuel when the full power of the engine is not needed. The changing of the position is known by most engineers as "hooking up."

The lead of the valve will be increased as the block approaches the center of the link. Thus a valve has more lead and also more compression when the engine is running with the valve hooked up. The link may be supported on a lifter rod, as shown in the cut, or it may be swung on a lifter from above. It may also be fitted with either one or two lifters; in the latter case one will be placed on each side of the link. However, the way the link is supported the action will be relatively the same.

Directions for Setting the Link Reverse.—With a direct valve and rocker or slide. First, proceed to locate the dead centers by means of a tram, as previously described, then remove the steam chest cover and loosen the eccentrics. Put the engine on its head end dead center and turn the forward motion eccentric with the point of greatest throw straight up at right angles to the crank and the back motion eccentric exactly opposite. Now place the valve in the exact center of the valve seat so that it will cover both steam ports the same amount. Make a light mark on the valve seat at the edge of the valve so that it may be readily placed in the same position should it be

VALVE GEARS AND VALVE SETTING

accidentally moved. Next put the reverse lever in the notch of the quadrant which will allow the link to be down and the forward eccentric rod in line with the valve stem. If a rocker arm is used it should be placed in a vertical position and the length of the valve rod so adjusted as to connect the valve and rocker arm when both are in their central position. If a slide is used in place of a rocker arm, place it in the center of its travel and adjust as stated above. Now adjust the length of the forward eccentric rod so that it will just connect the eccentric and rocker arm or slide when the eccentric is in the above position and the rocker arm central. Next throw the reverse lever in the last notch at the other end of the quadrant, raising the link and bringing the back motion eccentric rod in line with the valve rod. This eccentric rod should be adjusted so that it will also just connect the rocker arm or slide in its central position with the eccentric in the ninety-degree position already stated above. The above proceedings will give the correct length of the eccentric rods. Both adjustments should be tried, by moving the reverse lever from one end of the quadrant to the other, to see that they are correct in all respects.

Previous to adjusting this length of the eccentric rods, it is a good idea to test the length of the reach rod from the reverse lever, to the bell crank which raises and lowers the link. To do this place the reverse lever in the central notch of the quadrant and see if the link block is in the center of the link. In order to determine the center of the link measure the distance between the two pins at the top and bottom of the link and then make a mark on the latter halfway between these two pins. After adjusting the reach rod so that this mark is in line with the center of the link block pin when the reverse lever is in the center notch of the quadrant, securely fasten the eccentric, valve and reach rods in their respective positions.

In order to set the valve of the engine, place the reverse lever in the last notch of the quadrant at the end which depresses the link and loosen the forward motion eccentric which will be connected to the top eccentric rod. Rotate the eccentric the way the engine is to run in this motion ahead of the crank, enough to open the head end steam port about three-sixty-fourths of an inch. Temporarily secure the eccentric in this position, then turn the engine to the other center and see if the lead is the same. If not, move the eccentric ahead or back sufficiently to correct the difference. With the engine on its head center again, place the reverse lever in the last notch of the opposite end of the quadrant and turn the other eccentric ahead of the crank (remember that the engine is to run the opposite way from what it was with the other eccentric) sufficiently to open the head end port three-sixty-fourths of an inch. Now move the engine back to its center and see if the lead is the same; if not, adjust it by moving the eccentric slightly in the proper direction. After this the lead should be again examined on its first center as it is quite imperative that the lead should be the same on the two centers with both of the eccentrics. When this is all properly done and the lead adjusted for both motions, securely fasten both eccentrics; the job will be completed upon replacing the steam chest cover.

Some of the later types of engines using this style of reverse have the eccentrics keyed to the shaft or fastened by means of set screws which fit into the holes in the shaft. When this kind of engine is encountered, set the valve in the following way: Place the engine on its head end dead center, remove the steam chest and see that the reach rod from the reverse lever is of the correct length. Now disconnect the link, locate the valve central on its seat, place the rocker arm or valve slide in the central position and adjust the valve rod so as to connect them, when in their

VALVE GEARS AND VALVE SETTING

central positions. After doing this connect the link again and place the reverse lever in the end notch of the quadrant which will drop the link, thus connecting the forward motion eccentric with the valve rod. With the link in this position adjust this eccentric rod so that the valve will have about three-sixty-fourths of an inch lead on the end. Next turn the engine the way it is to run to the other center and examine the lead to see if it is the same. Should it not be the same, change the length of the eccentric rod one-half of the difference. When the lead has been properly divided, locate the engine on its head end center, throw the reverse lever in the last notch at the other end of the quadrant, and adjust the length of the back motion eccentric rod until the same amount of lead is shown at the head end steam port as was given with the other eccentric. Turn to the other center and examine the lead. If it is not the same divide the difference by changing the length of the eccentric rod. Be sure that everything is tightened before replacing the steam chest cover.

When setting any type of reverse gear the following points should be borne in mind:

1. Always turn the engine the way it would run if the steam were turned on. Should it be turned slightly past the center, either turn it quite a bit and bring it up slowly until the center is reached, or turn it completely over; this is to prevent any defect in the valve setting due to lost motion.

2. Always prior to setting a valve take up as much as possible all lost motion in every part of the engine.

3. Always have the reverse lever in the end notches of the quadrant and never in any intermediate ones between the end and the center notch, as to use these intermediate ones will prevent accurate work being done.

4. Always check over the work two or three times to make sure that it has been done correctly.

STEAM TRACTION ENGINEERING

fasten the eccentric. If, on the other hand, the engineer is not absolutely sure of the marks he should proceed to determine the position in the following manner. Put the engine on its head end dead center, turn the eccentric with the point of greatest throw almost opposite the crank pin and locate a position here so that when the reverse lever is thrown from one end of the quadrant to the other, the valve stem will not move. Several trials may be needed to locate this position but with perseverance it will eventually be determined. Secure the eccentric temporarily in this position and then determine the point of no valve stem movement by the same method with the engine on its crank end dead center. If the valve stem does not move when the engine is on either of its dead centers, then the eccentric is in its correct position and as such it should be securely fastened. If, however, the valve stem does move, the eccentric should be shifted until the amount of this movement is equally divided on both dead centers. This movement is caused by the pivoted guide not being of the correct height. It should be corrected.

To do this place the engine on one of its dead centers and move the reverse lever back and forth slowly, at the same time watch the guide and valve stem to see if the latter moves the same way as the top of the guide or in the opposite direction. If the valve stem moves in the same direction as the top of the guide, the latter is too low and should be raised slightly; but if the stem moves in the opposite direction from the top of the guide, then the latter is too high and must be lowered. The pedestal that carries the guide is attached to the engine frame and will be provided with liners placed between it and the frame. Should the guide be too low more liners need to be placed below the pedestal; but if, on the other hand, the guide is high, some of the liners already there should be removed.

VALVE GEARS AND VALVE SETTING

[a]fter changing the number of liners the pedestal should be [ma]d[e] solid and the height of the guide should again be [exa]mined. This should be continued until a position for [the ec]centric is found which will allow the reverse lever [to m]ove from one end of its throw to the other without [mov]ing the valve on both the engine dead centers.

[To] find the correct length of the reach rod from the [rever]se lever to the tilting guide, proceed as follows: [Thr]ow the reverse lever in one of its extreme end notches [of] the quadrant and then turn the engine over slowly, [car]efully measuring the distance the valve stem moves [fro]m one end of its travel to the other. Put the reverse [le]ver in the last notch in the other end of the quadrant, [an]d turn the engine over once more. If the distance the [va]lve stem moves is the same as that of the other end of [the] quadrant, the reach rod is of the correct length. If [not] the same, change it and repeat the above operation until [the] valve travel is equal.

Having found the correct position of the eccentric, the height of the pedestal and the length of the reach rod, remove the steam chest cover and disconnect the eccentric rod from the rocker arm. Next carefully measure and locate the valve centrally on its seat, then place the rocker arm in a vertical position and adjust the length of the valve stem so that it will connect the valve and rocker arm when both are in the above positions. After this adjust the nuts or blocks of the valve stem inside the steam chest so that the valve will have about one-sixty-fourth of an inch play between them. This will allow the valve to leave its seat without springing the valve stem. Place the engine on its head end dead center, throw the reverse lever in the "go ahead" notch of the quadrant and adjust the eccentric rod so that there will be about one-thirty-second of an inch lead at the head end. Turn the engine the way the reverse lever indicates and see how much lead

STEAM TRACTION ENGINEERING

is shown at the other end when the engine is on the crank end center. If the lead is more than one-thirty-second it is plain that the designer intended to give this particular engine more than the one-thirty-second-inch lead and as the lead is a fixed amount with this style of reverse, the amount of the lead will have to be divided equally. It is essential that the lead should be the same on either engine dead center when the reverse lever is in either the extreme forward or reverse position. If unable to obtain the proper lead adjustment the fault is no doubt due to an incorrect adjustment of some part of the reverse gear such as the height of the pedestal or the position of the eccentric. These should be done carefully over again to correct any possible mistakes which may have been made the first time. Before replacing the steam chest cover, make sure that all parts which were used in setting the valve have been properly tightened.

Having correctly set the valve, take a sharp cold chisel and mark the position of the eccentric if it has not been previously done. Where the eccentric is keyed to the shaft or otherwise definitely located, as by having holes in the shaft for the set screws, always center the engine and adjust the height of the pedestal carrying the guide, the length of the reach rod from the reverse lever and the length of the valve rod between valve and rocker as were stated above. Sometimes a worn-out block on the tilting guide will not allow perfect setting of the valve. This point should always be examined and if worn to any extent or out of true, it should be replaced with a new block.

The Springer Reverse Gear.—The improved Springer valve gear is illustrated in Fig. 83. At first glance it appears nearly identical with the Woolf valve gear previously described but it will be seen that the slide block and guide are slightly curved in this gear instead of straight as in the Woolf. This is the only material difference be-

VALVE GEARS AND VALVE SETTING

tween these two gears. Some of the minor details in either gear will be found different on the different makes of engines, but the general outlines of either type will be found the same and are well illustrated in the cuts.

The advantage claimed for the Springer over the Woolf gear lies in the shape of the guide and slide block. It is claimed that the curved block and guide of the Springer gives a quicker opening and closing of the ports than is

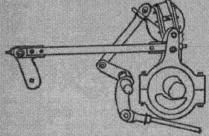

FIG. 83.—SPRINGER CURVED BLOCK REVERSE GEAR.

obtained with the straight block and guide of the Woolf gear. This causes the valve hardly to move until time to open and close the port, at which time it is done quicker than is possible with a straight block and guide, thus preventing any wire drawing of the steam. Another very important feature claimed for this valve gear is that it produces an equal lead and cut-off with the engine running in either direction and the reverse set at full stroke or "hooked up;" that is for any point of cut-off within the range of the valve gear.

Direction for Setting the Springer Valve Gear.—About all that need be said in regard to setting this valve gear is to follow the instructions given for setting the Woolf valve gear. As the two gears are almost identical and follow the same general principles, it of course operates in the same manner; therefore the same rules for setting it will also be used. The differences in the slide block and guide make no difference whatever in the setting of the two gears.

When setting the Springer valve gear proceed as with Woolf by first centering the engine, then determine the position of the eccentric nearly opposite the crank pin, the

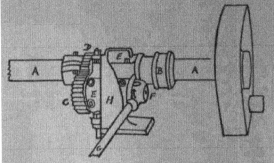

FIG. 84.—MARSH REVERSE.

height of the pedestal, the length of the reverse lever reach rod and the length of the valve stem. The lead of the valve is also adjusted the same as with the Woolf valve gear and is fixed by the designer; therefore it cannot be changed without some alterations being made in the valve gear.

The Marsh Valve Gear.—This valve gear illustrated in Fig. 84 is entirely different from those mentioned above, since it has no links or sliding blocks in its construction. The gear consists of a two part box EE, one end of which is mounted on the crank or main shaft, AA, while the

VALVE GEARS AND VALVE SETTING

other end carries a small crank shaft. The crank of this latter shaft, F, is directly connected to the valve by the eccentric rod, G. The opposite end of this small crank shaft has a small pinion, C, securely fastened to it; another pinion, D, of equal size is secured to the main shaft of the engine, and meshes into the pinion on the small crank shaft.

The governor pulley, B, is cast with a spool, the turned rim of which runs against the disk of the small crank. This spool serves to hold the box endwise against the gear on the main shaft. A stop plate, H, is secured to the engine frame to limit the travel of the box in reversing. The box rests against adjustable set-screw stops in this plate at either position of the box. The reverse lever quadrant is fitted with a center notch and an adjustable notch at each end of the quadrant.

This valve gear is very simple and with good care it should last for years with very few repairs. The cut-off cannot be varied with this reverse and therefore all this gear can do is to reverse the engine.

Directions for Setting the Marsh Reverse with Plain D Valve and Valve Slide.—On account of this gear being largely affected by expansion, it is necessary that there be about sixty or seventy pounds' steam pressure on the engine while setting this reverse. The dead centers should be properly located so that they may be turned to readily. If the exact position of the gear on the main shaft is plainly marked see that these marks coincide with each other. On the points of the teeth of the two gears will be found three plain marks, two of them being on one gear and one on the other. The marked tooth on one gear should mesh between the two marked teeth on the other gear in order that they be in their correct position.

If not sure of the position of the gear on the main shaft or if it is desirable that this position should be tested, the method to be followed is herewith given. Adjust the adjustable stops on the quadrant to about mid-position, i.e., about halfway between the two extremes of adjustment, and fix the length of the reverse reach rod so that when the reverse lever is in the middle notch of the quadrant, the reverse box will be exactly halfway between the extreme positions of the box. To find the center of the box travel, measure from the stop plate where the stop screws are located and not from the head of the screws unless they are of equal length. Now throw the reverse lever in one end notch of the quadrant and adjust the stop screw so that the box will rest against it snugly with the reverse latch tight against the stop on the reverse quadrant. Do the same with the other notch of the quadrant. Next turn the engine to the head end dead center and throw the reverse lever from one extreme to the other; while doing this watch the valve rod and see if it moves. Should it only move slightly and yet be in exactly the same position when the reverse lever is in either end of the quadrant, the gear is probably in its correct position. The small crank should be next to the main shaft when the engine is on the head end dead center.

In measuring the valve stem movement take a pair of dividers and placing one leg in a small prick punch mark on the valve rod stuffing box, make a light scratch on the valve rod with the free end when the reverse lever is in one of its extreme throws. Reverse the lever to the extreme end of the quadrant and measure again. If the valve rod moves or does not remain in the same position for the two positions of the reverse lever, shift the gear slightly on the main shaft and try again. Continue until a position is found that does not shift the valve rod when the reverse lever is moved. When this has been determined the gear should be lightly secured and the engine put on its crank end dead center. It will be noticed that in moving

VALVE GEARS AND VALVE SETTING

the reverse lever from one end of the quadrant to the other, the valve rod will move a considerable distance, but if the gear on the main shaft is in its correct position, it will be in the same position for either extreme position of the reverse lever. Place the reverse lever in one extreme notch of the quadrant and putting the dividers in the punch mark mentioned above, make a slight scratch on the valve rod with the free end. Now slowly move the reverse lever towards the other end of the quadrant, at the same time watching the lever rod. It will be seen that the valve rod will move towards the steam chest until the reverse lever is in the center of the quadrant, when as the lever passes the center it will move in the opposite direction until the end of the quadrant has been reached by the lever. The valve rod should now have returned to the same position as it occupied before moving the lever; if it should not so return the position of the gear will have to be slightly shifted on the shaft. A place can be found which gives the above position for both centers. The location of the gear can only be found by patient trials but once it is found the gear should be firmly secured and the place plainly marked.

After having found the correct position of the gears, remove the steam chest lid, center the valve and valve rod guide and then adjust the length of the valve rod to correspond. Now turn the engine to the head end dead center, drop the reverse lever in the notch of the quadrant which will allow the reverse box to be in its lower position and then disconnect the eccentric or valve rod. Adjust this rod so as to give about one-thirty-second of an inch lead on the head end, then turn the engine to run under to the other center and measure the lead; if it is not the same divide the difference on the eccentric rod. If after doing this the lead is found to be too much or if there should be no lead, the remedy is to change the set screw in the

STEAM TRACTION ENGINEERING

stop plate against which the box rests. For this direction of motion of the engine (running under) it is the lower set screw. After the proper adjustment of the lead (about one-thirty-second inch) has been made, screw the jam nut tight against the stop screw and firmly fasten the crank rod in its place. Now again place the engine on its head end dead center and throw the reverse lever over until the box rests against the upper set screw in the stop plate. Again examine the lead and if not right change the upper set screw. Turn the engine to the other dead center and measure it again; if there is a slight difference it is better to divide it by means of the set screw in the stop plate. Should the difference be a considerable amount some mistake has been made in the setting at some point and the only thing to be done is to go all over the work. Remember always to have the reverse box rest snugly against the stop screws in the stop plate when measuring the lead.

When the lead has been properly adjusted place the reverse lever in one end of the quadrant with the reverse box against the stop screw in the stop plate and then adjust the movable catch on the quadrant so that the reverse lever latch will just drop into it. The latch should be securely fastened in this position. Now throw the lever clear over so that the reverse box rests against the other screw in the stop plate and adjust the other quadrant catch in the same manner. After seeing that everything is fastened tightly, the job will be completed upon replacing the steam chest cover and turning on the steam.

An engineer may run across difficulty in setting this valve gear which may be entirely remedied by the rebabbitting the reverse box. If it should happen that this box is worn out of line as much as five-thirty-seconds of an inch, it would throw the lead adjustment out as much as five-sixteenths of an inch. When setting the valve gear of any

VALVE GEARS AND VALVE SETTING

engine it is well to examine first all the bearings to see whether they are badly worn.

The Arnold Reverse Gear.—The Arnold is a single eccentric reverse of simple construction having few wearing parts. The parts of this reverse are shown in Fig. 85, in which 1 is the main shaft of the engine; 2 the eccentric and strap; 3 the eccentric housing; 4 the shifting collar; 5 the shifting fork; 6 the eccentric rod, and 7 the reverse

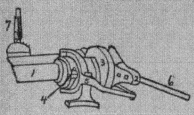

FIG. 85.—ARNOLD REVERSE.

lever. The plain eccentric is slotted so as to slide across the main shaft and it is cast with guides parallel to this slot which fit in slots in the eccentric housing. The housing is secured to the shaft and contains a slot of considerable length which has a set screw fitted at each end. The shifting collar slides endwise on the shaft by means of the shifting fork and reverse lever. The end motion of this collar is communicated to the eccentric through the eccentric housing and when the reverse lever is moved from one end of the quadrant to the other, the collar is shifted, thus causing the eccentric by means of the collar and connections to move from one end of the slot in the housing to the other. The set screws at the end of the housing slot regulate the distance through which the eccentric is thrown from the center of the shaft. It is obvious that with the

STEAM TRACTION ENGINEERING

lever in the center of the quadrant the eccentric will also be in the center of the slot. This slot in the housing is so cut that the angle of the advance of the eccentric is caused by this slot and is a fixed amount. This being the case when the eccentric is in the center of its travel in this slot, it will only have moved the valve the amount of the lap and lead (angle of advance). One eccentric only being used for both forward and backward motions of the engine, the housing must therefore have only one fixed position. The set screws in the housing are for equalizing and changing the cut-off. Lost motion in the reversing mechanism is easily and quickly taken up by means of adjustable notches on the quadrant. There is substantially no more wear on this reverse when the engine is in operation than on the plain single eccentric engine, the only wear of the reversing parts occurring when the engine is reversed.

Directions for Setting the Arnold Reverse with Plain D Valve and Valve Rod Slide.—While setting this reverse it is always better to have the engine hot because the setting is affected by expansion due to heat. Take the cover off the steam chest and see that as much as possible of the lost motion is removed. Locate the dead centers of the engine and proceed to find the correct position of the eccentric housing as follows: Turn the engine over to the head end dead center and the eccentric housing so that the eccentric slide is almost vertical on the opposite side of the shaft from the crank pin. This will cause the head end steam port to open. With the engine still on the head end of the quadrant take a pair of dividers and placing one leg in a small prick punch mark on the valve stem stuffing box, make a light scratch on the valve stem with the free end. Now slowly move the reverse lever to the other end of the quadrant and notice if the valve stem has moved. If it only moves a slight amount and is in the same position for each extreme position of the quadrant,

VALVE GEARS AND VALVE SETTING

the housing is probably in its correct place. On the other hand if the valve stem is not in the same position for the two extremes of the reverse lever, shift the housing slightly on the shaft and again try to see if the position of the valve stem will not remain the same for both positions of the lever. When such position has been found, lightly secure the housing and turn the engine over until it is on its crank end dead center. Again throw the reverse lever back and forth in the quadrant and see if the valve stem is in the same position when the reverse lever is at each end of the quadrant; if not it will be necessary to shift the housing until able to obtain the right position. If the valve stem is centered correctly a position can be found for the housing which will allow the valve stem to remain in the same position with the reverse lever in either end of the quadrant and when the other center is used the same results will be obtained without moving the housing. Remember that the two centers will give two positions of the valve stem, the difference in the two places being the amount of the lap and lead.

Having found the correct position of the housing, fasten it securely to the shaft. Next center the valve on the port face and place the valve rod slide in the center of its travel, then adjust the length of the valve stem so that it will connect them in their central positions. The eccentric rod should then be connected to the valve rod slide and the engine located on its head end dead center. The length of the eccentric rod should be so adjusted that the valve will have about one-thirty-second of an inch lead at the head end. After determining the lead on this dead center the engine should be placed on the other dead center (crank end) and the lead measured; if it is not the same it is evident that the builder intended a different amount of lead from the one-thirty-second of an inch. In this case the amount should be divided so that lead will be equal, at both ends of the valve. While doing this the reverse lever must be in the forward notch of the quadrant and the engine should be turned the way it runs when threshing when the reverse lever is in the forward notch. With the engine again on the head end dead center put the reverse lever in the other end of the quadrant and examine the lead. If the work has been correctly done there will be the same amount of lead as with the reverse lever in the other notch. As has been stated before, the lead must be the same when the reverse lever is in either extreme position of the quadrant and the engine is on each of its dead centers. When the lead has been properly adjusted for both motions, secure all the adjustments on the valve stem and eccentric rod.

The Arnold reverse, as before stated, is fitted with a set screw at each end of the eccentric housing which is for the purpose of changing and equalizing the point of cut-off. For ordinary work it is advisable to have the point of cut-off at three-quarters of the stroke; where the work is light and does not require the full power of the engine, the cut-off can be at five-eighths or one-half of the stroke of the engine. The directions for a cut-off at three-quarters of the stroke in both directions of motion of the engine are given herewith. First turn the engine from one dead center to the other, marking the position of the crosshead at each end of its travel on the guides and use the same end of the crosshead for both markings. Measure the distance between these two marks, which is the stroke of the engine. Suppose that this distance is ten inches, then three-quarters of the stroke would be three-quarters of ten or seven and one-half inches. When the stroke has been measured and the three-quarters of it ascertained, lay this distance off from both ends of the guides. Now place the reverse lever in the forward notch of the quadrant and turn the engine to the head end center. Move the

VALVE GEARS AND VALVE SETTING

engine slowly in the direction of the belt motion until the crosshead comes to the mark on the guides made for the three-quarter cut-off. This will be the cut-off mark farthest from the cylinder. Loosen the catch on the quadrant and change the set screw against which the eccentric rests in the housing, until the head end port is just closed. Set the jam nut on the set screw, and change the movable catch on the quadrant so that the lever holds the eccentric against the set screw. Now turn back to the head end

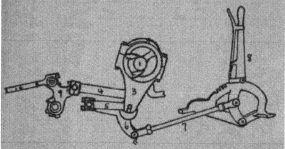

FIG. 86.—REEVES NEW REVERSE AND EXPANDING GEAR.

dead center, place the reverse lever in the back notch of the quadrant with the catch loosened and turn the engine in the back motion to the cut-off mark nearest the crank. Adjust this screw until the port is just closed, then secure the jam nut and adjust the movable stop on the quadrant. To test these settings, turn the engine to the other cut-off mark and move the reverse lever from one notch to the other, examining the settings at each position of the lever. For other points of cut-off merely change the marks on the guides to the proper place and proceed as above.

The Reeves Reverse Gear.—The Reeves reverse and expansion gear, as it is called by the builders, is illustrated in Fig. 86. It is claimed to be very durable and accurate, giving a perfect motion to the valve at all points of the cut-off with the engine running in either direction. It contains no links, rollers or sliding blocks in its construction, having but one eccentric and rod with two arms. In the cut, 1 is the main engine shaft; 2 the eccentric; 3 the eccentric strap; 4 the eccentric rod; 5 the radial rod; 6 the reversing arm; 7 the reach rod; 8 the reverse lever; 9 the rocker arm, and 10 the valve stem.

The principle of design and of operation of this reverse is similar to the Springer gear above described. The motion of the eccentric controls the lap and lead of the valve, any further movement being obtained by the angle at which the reverse arm and the radial rod stand in relation to each other. When in the central position as shown in the cut, all movement of the valve is obtained from the eccentric but when the reverse lever is thrown either forward or backward, it places the reverse arm and the radial rod at an angle with each other; the further the lever is moved the greater this angle becomes. Thus the travel of the valve, and hence the point of cut-off, is varied according to the position of the reverse lever from the center of the quadrant; the further from the center the lever is placed the later the cut-off. The lead remains constant at all points because of the eccentric giving only the lap and lead to the valve. As only one eccentric is used for both forward and backward motions of the engine, its position must be a fixed one and the lead must be a fixed amount; therefore if the lead should be changed so as to become greater on one motion, it would be less on the other, which is not advisable for smooth running or economical operation of the engine.

Directions for Setting the Reeves Valve Gear with Plain Slide Valve.--Remove the steam chest cover and locate the

VALVE GEARS AND VALVE SETTING

centers of the engine. Take up all lost motion as far as possible and place the engine on the head end dead center. The position of the eccentric is nearly opposite the crank pin. To locate it exactly, leave the engine on the head end dead center and move the reverse lever from one end of the quadrant to the other, meanwhile watching the valve stem; if it moves when the reverse lever is moved set the eccentric slightly in the shaft. Continue until a position is found which will allow the reverse lever to be moved from one end of the quadrant to the other without moving the valve stem. When such a position has been found, secure the eccentric lightly and turn the engine over to its crank end dead center. With the engine in this position throw the reverse lever back and forth, meanwhile watching the valve stem. If it does not move, the eccentric is in the right position, but if it moves slightly a slight mistake may have been made in centering the engine. If a position cannot be found for the eccentric that gives no movement of the valve stem on either center, the reverse arm is probably too low or the radial arm too short. These two defects are caused by wear and will only occur in engines which have been run some time. The reverse arm may be raised temporarily by placing a piece of cardboard under the supporting shaft in its box. If this does not correct the trouble place a piece of cardboard in the radial rod bearing so as to lengthen the rod. If this should help, continue to add cardboard until both centers of the engine have been corrected. All that will be needed is about two thicknesses of cardboard as more wear than this is not likely to occur. When only a slight movement of the valve stem is encountered, it is best to divide the amount as changing the reverse arm or radial rod is a delicate operation. Having located the eccentric fasten it securely to the shaft and plainly mark the position of the shaft and hub of the eccentric.

The method to be followed in order to adjust the length of the reach rod from the reverse lever, is to place the lever in the forward notch of the quadrant and by turning over the engine to measure carefully the valve travel on the valve stem. After the above measurement has been taken, place the lever in the back notch of the quadrant and measure the valve travel, while the engine is being turned. If the valve travel is the same for both positions of the reverse lever, the reach rod is of the correct length, but if they should be different, the length of the reach rod must be changed. To cause more travel on the forward motion, lengthen the rod and vice versa. The rod should be changed until both motions are of the same length.

When this has been adjusted, turn the engine to the head end dead center, place the reverse lever in the forward notch of the quadrant and give the valve about one-thirty-second of an inch lead on the head end. This must be done on the valve stem. After this has been done turn the engine the way it would run when at work, to the crank end dead center, and examine the lead. If it should not be the same, divide the difference until the lead is an equal amount on both ends of the valve. Place the reverse lever in the back notch of the quadrant, turn the engine in the reverse way, and examine the lead. If only a very slight difference is noticeable, nothing need be done about it; but if, on the other hand, the amount is considerable, it should be corrected. Remember that if the eccentric is in the proper position, the reverse arm at the proper height and the radial rod of the proper length, the lead will be exactly the same for both forward and backward motions. If there is any trouble in getting equal lead at all points, the previous work should be gone over again to make sure that no mistakes have been made.

When the correct position of the reverse arm and the length of the radial rod have been determined, if they were

VALVE GEARS AND VALVE SETTING

previously wrong, they had better be rebabbitted in their new positions. Before replacing the steam chest cover see that the jam nuts and set screws are properly tightened.

Other Styles of Valve Gears.—An engineer may run across other styles of valve gears while following the calling of farm engine operator, but the instructions given for the gears previously mentioned will cover as well all the basic principles of other gears. The engineer's mechanical ability should teach him how to modify the knowledge which he has gained, or should have, from the instructions already given so that it will fit the special gear under consideration.

Mention might be made of two reverse gears which have not already been previously given, namely, the Russel and the Peerless. The Russel gear works on the same principle as the Arnold reverse, with the additional distinctive feature in that it is able to change its cut-off by "hooking up" the reverse lever. The Arnold has a fixed cut-off which cannot be changed without considerable work and measurements.

The Peerless gear is made into styles both of which have the same basic principles as the Arnold and Russel gears. All single eccentric reverse gears have a fixed lead. By this is meant that the amount of lead has been worked out by the designer and the gear is proportioned to give this lead. The lead on a valve with this kind of reverse can be changed in only two ways, i.e., by changing the valve or by a change in the reverse gear. As either change would entail new parts the cost would be prohibitive and the advantage to be obtained would only be probable, not material. Remember that the designer is one who has had years of experience and that the engine has been thoroughly tested by the builders with this amount of lap and lead, before being placed in the market. Therefore do not come to the conclusion that the engine is not designed

STEAM TRACTION ENGINEERING

commonly or that the builders made a mistake and that
operating engineer can make an improvement in the
ciency of the engine by changing the amount of the
of advance because he will soon find out that it canno
them.

CHAPTER VII

REPAIRING OF ENGINES

An engineer should be able to do a certain amount of engine repairing as in the case of boilers. There are certain kinds of work about an engine which may require the attention of an expert mechanic and a well-equipped shop; but on the other hand all minor repair work which is required to keep the machine in good running condition should be done by the operating engineer. It is not advisable for an engineer to do a certain piece of repair work unless he is reasonably sure that he knows what is to be done and how.

Adjusting Crank and Crosshead Pin Bearings.—Nearly all of the latest design of crank pin bearings are adjusted by means of a wedge-shaped block which is moved up or down by a threaded bolt or two tap screws. When the threaded bolt is used, a jam nut is used on the lower end to prevent the bolt from turning or working loose. Should the tap screws be used, one of them will be found to screw into the block from above and the other from below. The box is in two halves and should have about one-eighth to one-quarter of an inch space between the ends to allow for wear. When tightening the box loosen the lower set screw or jam nut and screw the upper bolt into the block. This draws the block up and tightens the box by bringing the two halves closer together. Drawing the wedge down will of course loosen the box. When the proper adjustment is secured tighten the jam nut or lower set screw. The crank pin should always be adjusted when the engine

TEAM TRACTION

h   nter as this is the p
pin.   In order to deter
engin er will have to u
   careful not to get th
   a few minutes nor must
nocks at every stroke. Where the
f end play on the p   good way
ill just slip back and forth the
y taking hold of the c nnecting
asly. Where the box s adjusted
ery good way to obtain a running
ie key down tight, then loosen it
ie hand. This will nearly alway
iat will not heat. It is well to r
ous pressure can be exerted on a
r the wedge block. A great many
ell as some older ones, make the
ike all the knock out of an engin
rank pin box. It is a peculiar thing
xplained, that a knock in almost any
ill appear to be in the crank box.

As stated above, the brasses are usua
f an inch apart. When this amount of
nd has been taken up, the boxes will be
) allow for further adjustment abou
n inch is filed off each of the brasses.
ave worn down the one-eighth of an i
ie length of the connecting rod between
ie crosshead pin centers to become
iorter by just one-half of the amount
epends on which side of the box the a
s to which way the length of the rod
f nothing was done to overcome this
short time before the length of the r
ifficiently to cause the piston to str

REPAIRING OF ENGINES

Also some means must be provided to keep the outside of the box the same length as it was originally, so that the key or wedge will be able to take up the slack. Since the box has worn one-eighth of an inch it must necessarily be one-eighth of an inch shorter on the outside. To replace or fill this vacant space, backing or shims, as they are called, are used. Sheet copper is one of the best materials to be used for this purpose. Heavy sheet iron can also be used. There are also laminated metal shims. The metal should be cut so that it will cover the whole end of the box where it fits against the shoulder of the rod, wedge or key. In placing these shims in either place, put half on each side of the box. If one-eighth inch has been worn off, place one-sixteenth inch backing on each side of the box. This will correct the difference in the rod length and keep it even, at the same time keeping the box long enough for further adjustment.

When filing down the edges of the half-brasses, always dress the inside edge tapering. The taper should be about one-sixty-fourth of an inch deep at the edge and should extend back about one-quarter or one-third of an inch. This prevents the brasses from pinching the pin and will make a much smoother running bearing. Another reason why brasses should be relieved on the ends where they come together is the pinching caused by the wear of the brasses. This is brought about in the following manner: As the pin wears down the brasses the tendency is to wear faster in line with the push and pull of the connecting rod, than up and down, and this is the simple reason why the brasses will tend to pinch the pin at the top and bottom. This condition is usually called a "brass-bound" crank box and will be found very hard to keep from heating if kept tight enough to prevent knocking.

The crosshead pin brasses are about identical with the crank pin brasses and therefore should be handled in the placed between the halves to allow the nuts to be screwed down tight. Remember that a slight turn too much on the set screws holding the gibs of the quarter boxes will cause them to heat. Take these boxes up a very little at a time until the slack has been removed and then leave them alone. It is not necessary to tinker with these bearings every day to keep them in good shape. If a bearing is once set properly it will not require any further adjustment for a considerable period. Some of the later types of split boxes have a small chain or ring around the shaft which runs in a large groove and oil chamber. These oil chambers and grooves with the chain or ring should be cleaned out once in a while and entirely fresh oil supplied. Quarter boxes, when made of brass, may be babbitted when worn out or else renewed (page 251). Where the quarters are made of cast blocks with babbitted faces, the renewal can be accomplished by the same means, that is either by rebabbitting them or getting new blocks.

Adjusting the Crosshead.—Should the crosshead wear loose in its guides it will cause a very undesirable knock or pound. The method to be followed in order to eliminate the trouble will depend entirely upon the style of crosshead and guides. In the locomotive pattern only the guides are adjustable. There are usually a few layers of shims between the guide bars and the lugs on the engine frame to which they are bolted. Enough of these shims should be removed to take up the play and still not bind the crosshead between the guide bars. Always have the crosshead at the end of its travel when adjusting the guide bars. If the engine has been in use quite a long time all of the shims will have been removed. In that case remove the bars and carefully file them down sufficiently to remove the play in the crosshead. Be careful while filing to get the bars perfectly true and square with the original bearing or else the crosshead and bars are apt to

REPAIRING OF ENGINES

start cutting and will soon be ruined. The most accurate and satisfactory way is to take the bars to a machine shop and have a small amount planed off each of them. If this work is required to be done during the busy season when the engine is working in the field the engineer will probably have to file them instead of sending them to the shop to be fixed. They should be adjusted so that the crosshead travels in a true straight line and does not allow the piston rod to "drag" the stuffing box. The side play of the crosshead can be prevented by adjusting the guide bars sidewise sufficiently to take care of all lost motion. If the bolts holding the bars will not allow them to be moved the necessary amount, file the holes with a round file enough to allow them to remove the amount of the play. It is a very good plan to wedge them in this position with a small piece of lead or babbitt driven between the bolt and the bar on the opposite side from the crosshead. Place sufficient shimming under the bars to allow the holding down bolts to be drawn up tightly.

With crosshead and guides of the Corliss pattern (Figs. 40 and 41) the matter of taking up the slack or looseness in them is a much easier task than with the locomotive pattern mentioned above. In the Corliss pattern the wear is taken up on the crosshead, the guides remaining stationary. By this arrangement there is no liability of the guides getting out of alignment with the cylinder, which is likely to happen with any engine made with adjustable guides. When any looseness appears in the crosshead with this type of engine, the shoes on the crosshead are set out sufficiently to compensate for the amount of wear that has occurred. This is accomplished in several different ways. One of the most practical and universally used methods is by means of wedge-shaped shoes, which slide endwise on a wedge-shaped seat on the crosshead. Sliding these shoes endwise on the crosshead increases or decreases the amount of the crosshead, according to which direction the shoes are moved, and causes it to fit the radius of the guides either tightly, loosely, or any degree desired between these two points. The shoes are usually held by means of clamp bolts. When an adjustment has to be made on one of these crossheads, turn the engine over to one of its centers (it does not have to be the true one) so that the crosshead will be at the end of its guide. Then loosen the nuts holding the movable shoes, slide the shoes endwise until the slack or looseness is removed and then tighten the bolts. In making this adjustment do not get the shoes out too far thereby making them too tight, or else the guides will become hot and begin to cut, causing a big damage to the engine. In making this adjustment of the crosshead be sure that the piston rod runs in the center of the cylinder and does not drag when the crosshead is nearest the cylinder.

Another style of the Corliss pattern crosshead which is used to quite an extent makes use of set screws to adjust the shoes. They are fitted with a holding bolt and an adjusting screw at each end of each shoe. To set the shoes out, the holding bolts should be loosened and the shoe set out the proper amount by means of the adjusting screws. Be sure to tighten the bolts again after making the adjustment. In all other respects they are handled in the same manner as the style mentioned above. Care should be taken with this style to see that one end of the shoe is not set out farther than the other, in order to prevent the crosshead from running out of line. When the shoes become worn until the adjusting screws no longer set them out, they will have to be removed and shims placed between the shoes and the crosshead and adjusting screws. With the style having sliding shoes as first mentioned when the shoes become worn until they will not adjust outward any farther they may be pushed back the other

REPAIRING OF ENGINES

way as far as they will go and a few layers of sheet copper or galvanized iron placed between the shoes and the crosshead. When this has been done the shoes will be able to be adjusted as before.

A few engines are built with the Corliss pattern guides which do not have any means provided to adjust the shoes other than shimming. With this kind of crosshead the shoes have to be removed and a piece of copper or tin of the proper thickness used between the crosshead and the shoe to hold it out sufficiently to prevent any lost motion. To do this it will be necessary to disconnect the engine either at the crosshead or at the crank pin. This will allow the crosshead to turn or twist sidewise until the shoes can be lifted off. Never disconnect the engine at the crank pin or the crosshead with steam on unless there is a tight stop valve in the steam pipe between the throttle and the boiler. Otherwise if the throttle should leak or be accidentally opened the piston may be driven through the cylinder with disastrous consequences. Some engines have the piston rod screwed into the crosshead and further secured by a jam nut or a clamp. Always be sure that the piston rod is securely fastened to the crosshead. The jam nut or clamp should be frequently examined to see that it is tight. Where the piston rod is secured by means of a key, the same care should be exercised in keeping it tight. If the key works loose a few times trouble is sure to result until a new one is fitted. A badly wrecked engine will be the result of the piston rod becoming disengaged from its crosshead while the engine is running. For this reason the engineer should give particular care to keep the piston rod tight on the crosshead.

Adjusting Eccentrics.—An eccentric yoke requires very careful adjustment in order not to get it too tight. The heating of the eccentric and its strap or yoke will cause them to stick, with the result that the eccentric is very

STEAM TRACTION ENGINEERING

apt to slip on the shaft. When the eccentric is key[ed] to the shaft, the strap or yoke may become brok[en] or at least badly cut, so that it will be hard to get ba[ck] in good running shape. A loose eccentric yoke, or stra[p] if not kept tight, will make quite a bit of noise, as w[ell] as fail to give the valve the exact movement. This yo[ke] is made in halves with packing between them and he[ld] together with bolts. To take up the lost motion in t[he] yoke, withdraw the bolts, extract a small amount of t[he] liners, being careful to remove as nearly as possible t[he] same amount from each side, and screw the bolts tig[ht] again. Remove only a small amount of the liners at a ti[me] as it is better to have to take up on the eccentric thr[ee] or four times than to have it get hot once. When the line[rs] have all been removed, the yoke should be taken to [a] machine shop and a small amount planed off the ends [of] each half. Liners should then be put in until a runni[ng] fit is obtained, which may be withdrawn as further we[ar] occurs. If not close to a machine shop or if it is not co[n]venient to take them there they may be filed sufficient[ly] to take care of the wear. In filing be careful that th[ey] are made as true and square as possible, so that the halv[es] will not bind on the eccentric when tightened by the bolt[s]. Eccentric yokes are sometimes babbitted, in which ca[se] they should be rebabbitted instead of being planed [or] filed down.

Fitting a New Crosshead or Slides.—This is a job whic[h] an engineer will not be required to do very often but may become necessary at times. Should the crosshead [or] slides become broken or worn to such an extent that the[y] cannot be adjusted with any degree of satisfaction, the[n] they will have to be renewed.

Since the crosshead of the locomotive type is solid, [it] is not capable of adjustment and therefore when it be[comes] badly worn it must either be replaced with a ne[w]

REPAIRING OF ENGINES

one or else be rebabbitted (page 257). The latter method is only practical where the crosshead pin is in such shape that it can be re-dressed and the piston fastened in such a condition that it is perfectly secure. The only part that can be babbitted is the slide bearing. When refitting a new crosshead of this type, it is best to be sure that the piston rod and crosshead are securely fastened to one another and that the former is in perfect line with the latter. If it is secured with a key, the rod must fit the hole in the crosshead snugly and the key must be firmly and accurately fitted. If the key should be made slightly too small, do not try to bush it to a fit, but make a new one. The new key should have exactly the same taper as the old one and be of a sufficient thickness to fill the key seat. After properly securing the crosshead to the piston, arrange the guide bars the proper distance apart and in exact line with the center line, or travel of the piston of the engine. This is absolutely essential in order to obtain a smooth-running engine. The bars should be adjusted to a running fit, both up and down as well as sidewise. Watch the slide bearings very closely for a few days and keep them plentifully supplied with oil until they have had a chance to wear smooth. If this precaution is neglected and the bearings are a little rough, the crosshead may start cutting in a short space of time. If by any chance one of the guide bars becomes sprung or broken, it should be replaced with a new one. It is sometimes advisable to put in a whole new set as trouble is very liable to be encountered again while adjusting the bars. Just as much care must be taken in adjusting the new bars as is taken when replacing a crosshead. They should be watched closely until they are worn down to a bearing in order to avoid all tendency toward heating or cutting. The guide bars wear faster in the middle than at the ends, due to the increased pressure of the crosshead at this point,

STEAM TRACTION ENGINEERING

caused by the greater angularity of the connecting ro[d]
After considerable use, if the bars are adjusted to a prope[r]
fit with the crosshead while the latter is at one end of i[ts]
stroke, there will be considerable play in the center [of]
its travel. It also stands to reason that should the ba[rs]
be set to take up the play in the center, the crosshead woul[d]
have to wedge its way between the bars at either end of t[he]
stroke. This would either spring the bars or break som[e]
thing; hence the instructions about always adjusting th[e]
bars or the crosshead when the latter is at one end of th[e]
stroke. The engine will not run as smoothly when thi[s]
condition is present as when the crosshead fits snugly th[e]
whole length of the guides. The remedy for this is t[o]
take the bars to a machine shop and have them plane[d]
down until they are perfectly straight. Never try to dre[ss]
the guides down by hand as the job can never be don[e]
satisfactorily.

With the Corliss type of crosshead the body will la[st]
as long as the engine, unless an accident occurs, resultin[g]
in the breakage of the body, or the crosshead pin work[s]
loose for a considerable time, which wears the pin holes i[n]
the body beyond repair. When this condition occurs th[e]
body will have to be replaced with a new piece. All that i[s]
required when fitting a new body is to secure it firmly t[o]
the piston rod; then replace the pin and shoes and properl[y]
adjust them according to the directions given under th[e]
heading of Adjustments.

When a shoe has been broken about the only remedy i[s]
to replace it with a new one. This is a short and simpl[e]
job as all that is necessary is to remove the broken sho[e]
and put the new one in its place. The shoe should b[e]
accurately adjusted and watched closely until it is wor[n]
to a good bearing in order to prevent the possibility o[f]
its heating or cutting. If the shoes are worn down unti[l]
there is considerable backing required to adjust them

REPAIRING OF ENGINES

It will be found a little difficult to hold them securely. To remedy this the old shoes may be replaced with new ones or they may be babbitted. When replacing the new ones, care should be taken not to get them too tight in the guides or they will begin to heat and cut. They also should be so adjusted that the piston rod travels in its central position and does not bind in the stuffing box. New shoes will require considerably more oil for a few days until they are worn smooth. Old shoes can be repaired by babbitting if they are worn as much as one-quarter of an inch. The directions for doing this job are given on page 258. When adjusting the shoes, have the crosshead at one end of its stroke on account of its tendency to wear faster in the center of the crosshead travel than at the ends. This wear is much slower with the Corliss type of guides than with the locomotive type, for the reason of the much larger bearing or wearing surface. On this account they will become more concave after long use and heavy work, so that it will almost be impossible to adjust the crosshead so that it will run nicely. This class of guides, being rigid with the engine frame, are not very easily repaired when this condition occurs. The only remedy is to have the guides re-bored in a well-equipped machine shop. After re-boring the guides the crosshead shoes had better be babbitted as the guides will be quite a bit farther apart. The new shoes in the beginning will require shims. After the guides have been re-bored and the crosshead refitted they should be watched and kept well supplied with oil until they have been worn smooth. The crosshead and its various parts including the guides should be examined and supplied with oil at all times, but during a short period of days immediately following the repair or replacement of any of the parts, the engineer should give an extra amount of both watchfulness and oil.

The only safe remedy in case of breakage of one of the

guides is to get a new girder. With the Corliss type of guides the girder usually comprises the engine frame, main pillow, block and guides and sometimes the cylinder and steam chest. This is a pretty expensive piece. It must be attached to the brackets, the main or crank shaft bearing babbitted and the crosshead fitted to it. If only a portion of the guide has been broken off it can sometimes be repaired by taking a couple of heavy iron straps and fastening them by means of tap screws. These screws should just pass through the holes in the straps snugly and be screwed into the threaded holes of the casting tightly. They should not pass through the guide or a rough place will be made on the bearing surface of the guide which cannot be smoothly dressed.

These straps must be heavy and the holes must be drilled very carefully so that when the screws are screwed in, the broken piece will be held tightly against the other and also held in perfect alignment. Patching a broken guide is a pretty particular piece of work which must be done accurately and will seldom prove satisfactory in the end, especially on heavy engines.

Fitting Crosshead Pin and Brasses.—After an engine has been run for a considerable length of time, difficulty will be experienced in getting the crosshead pin brasses tight enough to prevent the box from knocking without causing the journal to heat. This is due to the pin becoming worn oblong or flat, since almost the entire wear on this journal is in line with the piston, due to the bearing only oscillating slightly instead of turning completely around. It is evident that the remedy is to dress the pin until it is made round. If the pin is removable, as is the case with the Corliss type of crosshead, it should be taken out in order to dress it. Where the pin is not removable, as with the locomotive type of crosshead, it will be required to be dressed in the crosshead. If a machine shop

REPAIRING OF ENGINES

is near the place where the engine is working and the pin can be removed, it is best to have the pin taken to it for dressing. If the shop is too far away the engineer will be able to dress it by following the instructions given below:

Adjust a pair of calipers so that they will just fit the pin at the point of smallest diameter of the bearing surface. File down the large diameter of the pin until the calipers will just lack a fraction of slipping over the pin. After the pin has been worked down to this point, finish the remainder by first "draw filing," i.e., use the file like a knife by holding it flat to the work and slowly working it back and forth, until the calipers fit the pin the same at all points of its length and circumference. Finish the job by using a strip of emery paper on the pin to remove all the roughness. The job of dressing a pin will be hard when it cannot be taken out of the crosshead but it can be done, provided the engineer has perseverance. An engineer must use his judgment as to when a pin should be dressed or when it is better to purchase a new pin and crosshead body. Bushing a removable pin which has worn in the crosshead holes until it cannot be kept tight in the crosshead, may make it hold a while but it is seldom found to be a satisfactory repair job. The only correct remedy in this case is to get a new pin and crosshead body. A new body will be needed, for the reason that the holes for the pin will be worn fully as much as the pin. Where the pin is held by means of clamps, the dressing can be avoided for a long time by turning the pin about one-quarter of a turn every time it is removed to refit the brasses. This will present a new wearing surface to the brasses and also prevents the pin from wearing flat. Sometimes there will be encountered a pin in which particles of the brasses have become imbedded due to excess heat of the pin and brasses, caused usually by too tight a

STEAM TRACTION ENGINEERING

bea[ring for] lack of oil. This makes the pin and brasses very rough a[n]d it is a difficult matter to get the box to run without heating or pounding. When a bearing in this shape is encountered, remove the brasses and if the trouble is found to be in the pin, make it smooth again, either by draw filing it first and finishing it off with emery or using the emery cloth only, according to how badly the pin is cut. To smooth the bra[sses they] should be scraped. When using a scraper on any[thin]g do not make it cut too fast but instead work slo[w an]d carefully, removing only enough to take out all the ro[ughn]ess in the bearing. Scrape in the same manner as wh[en d]raw filing. A scraper can be made from an old flat [file b]y grinding off the teeth so that it is perfectly smooth and then putting a sharp edge on it. A good scraper is sharpened to a point and the end slightly curved.

When putting in a new crosshead pin or refitting an old one, if the bearing is oiled through the pin, as they nearly all are, see to it that the hole in the center of the pin and the small hole leading from it to the wearing surface are not stopped in any way with dirt, chips or particles of metal. Trouble is liable to happen if these parts do not receive perfect lubrication. Whenever the pin on a Corliss crosshead held by means of tapering shoulders or offsets and drawn up by a nut is removed, be sure that the pin is well drawn up so that the brasses will not be worn out of shape by the shoulder or taper of the pin. This is an important point that should be attended to. If the pin will not go up to its place there is something wrong which needs attention. In nine times out of ten the trouble is due to the pin having become battered a little and it can be repaired by being dressed. Sometimes a little dirt or chip in the hole will prevent it from fitting. The holes should be wiped clean with a piece of waste or rag before putting the parts together. Never dress a pin un-

REPAIRING OF ENGINES

less it is imperative or else it will begin to work in the crosshead and once it begins to work a new one will shortly be required. The pin should at all times fit the crosshead tight.

When a new crosshead pin is required and must be fitted all there is to be done is to dress the brasses to fit the pin and place them in position. Remove the old pin first and then carefully fit the brasses to the new one. If they should bind on the pin they should be carefully scraped until they give a proper fit. When the brasses have been fitted they should be relieved on their top and bottom edges so that there is sufficient space left between the halves to allow for take-up in the case of future wear. The instructions for relieving are given on page 227. When the brasses have been worn thin or a great amount of shimming is necessary, the best and quickest remedy is to replace the worn ones with new parts. The old brasses can be babbitted instead and should the job be carefully done they will prove satisfactory when repaired in this manner (page 253). When new brasses are to be fitted it is advisable either to put in a new pin or else dress the old one. This will not leave a rough journal to start with and will be much less liable to heat or cut. Brasses which are new should be cut down so that ample space is left between them for taking up the slack due to wear, and the edges also should be relieved. Should they become loose in the strap they will cause a very disagreeable knock and tightening the brasses on the pin will only cause the knock to be louder and more troublesome. The knock can be removed by making a shim out of tin or sheet iron of a sufficient thickness to fill the space between the brasses and the strap. It is best to place this shim under the box as then it will not interfere with the oil hole. The brasses should always fit the strap so tight that they need to be lightly tapped into position. A block of wood

STEAM TRACTION ENGINEERING

should always be used when driving brasses in and out of the strap or when removing the crosshead pin.

Fitting Crank Pins and Brasses.—Since the crank pin bearing travels at a medium high velocity it has heavy jars to sustain and usually a heavy frictional load. On this account it is more apt to give way and will require therefore more attention than any other bearing on the engine. Unless kept properly adjusted and well oiled it will get hot and start cutting in a very few minutes. Should this bearing give trouble it should be fixed at once.

When a pin has been somewhat worn it is apt to become rough and become cut in ridges. The rough places should be made smooth by dressing the parts by draw filing and emery cloth. This dressing must be done without removing the pin. If the pin has become brassed by getting hot, the roughness should be removed in the same manner. Instructions for this work will be found on page 227. If a new pin is required on a center crank engine it will mean a main shaft, as the crank pin or pins and shaft are all in one forging. The crank pins on nearly all side or disk crank engines are set in the disk by means of a hydraulic press and are usually riveted over, although sometimes they are secured by means of a nut. To remove the old pin the nut is withdrawn and the pin ejected by using a heavy punch and hammer, or a jack. In the case of a pin which is riveted, the burr caused by the riveting must be cut off by a chisel in such a manner that the end of the pin on this side of the disk is left beveled. If the pin cannot be removed in this manner it will have to be drilled out. This will be a tedious job which will have to be done carefully in order to avoid drilling into the crank disk and leaving the pinhole out of shape, thus spoiling the job. After the old pin has been removed see that no chips or dirt adhere to the inside of the hole. The

REPAIRING OF ENGINES

new pin should protrude out from the face of the disk about a quarter to three-eighths of an inch when shoved into the hole by hand. If it is a little too large to do this, it may be dressed the proper amount with emery cloth. When dressed to the right size it is ready to be driven in place, which is done by first heating the disk around the hole hot enough to melt wire solder. Next lay the torch aside and quickly place the pin in the hole and drive it home by placing a block of wood on the end of the pin and hitting it with a hammer. As soon as the pin has been driven up do not strike any more blows as they would tend to loosen it in the hole. The contraction of the hole on cooling tends to grip the pin. If the pin is to be riveted it should not stick through the disk more than one-sixteenth of an inch. With someone holding a heavy block of wood against the outer end of the pin, rivet it with light blows of a hammer. When a nut is used, it must be driven home tightly and the end of the pin cut off flush with the nut, then slightly "swelled" with a ball pein hammer. If the engineer is not sure that he can do this job with certainty he had better have it done in a machine shop.

Sometimes a badly damaged crank pin on a center crank engine may be turned down at a machine shop and the brasses then babbitted to fit the pin as the turning down will make it quite a bit smaller than before. Crank pin brasses will be ruined if allowed to get hot and stick to the pin. This causes the "brassed pin" which has already been mentioned and also causes more or less damage to the brasses as well. Small particles of brass are torn from the bearing surface of the brasses and stick to the pin. These rough projections on the pin result in the brasses being badly scored and gouged in a few minutes of engine running. In a case like this both the pin must be dressed and the brasses scraped in order to avoid any future trouble. These brasses should be refitted in the same manner as the brasses of the crosshead pin. Whenever a new crank pin is put in or the old one dressed it is always best to scrape the brasses to a nice snug fit on the pin. If a new set of brasses is to be put in, the pin should always be dressed and then the brasses should be carefully fitted to it. In all cases whether refitting old brasses or new ones, they should be relieved on both top and bottom edges and the ends dressed to allow sufficient room for taking up future lost motion. If the brasses should be loose in the strap they will make a very disagreeable knock which should be stopped before they have time to wear to any extent. They should be shimmed in the strap in the same manner as that described for crosshead pin brasses. They should always fit the strap just tightly enough to require being driven in or out with a few light blows of a hammer. In order not to bruise the brasses a block of wood should be used.

New brasses will have to be fitted or the old ones babbitted when they have become worn to such an extent that they are too thin for any further use or they require a large amount of shimming to bring them up to the pin. These brasses also wear on their sides and in time will wear enough to allow considerable end play on the pin. On account of this a knock will appear if the engine should become out of line in the slightest degree. This side play may be taken up by removing part or all of the liners between the cap on the end of the pin and the pin itself. If there is still some side play after removing all these liners, two liners cut from heavy sheet copper should be placed on the pin. One of them should be placed next to the disk and the other on the end of the pin. Should these not be sufficient more washers may be used. When a washer like this is made it should be dressed smooth on both sides, should fit the pin snugly and ought to be the same width as the bearing ring of the cap on the end of the pin. When making the washer it may be advisable to have

REPAIRING OF ENGINES

it thicker than necessary and place rings of cardboard or paper between the cap and the pin until it is adjusted to its proper distance. By this arrangement means will be provided for taking up future wear without having to resort to using more washers.

FIG. 87.—GROOVING BRASSES.

Cutting Oil Grooves in Crank and Crosshead Pin Brasses.—Upon the proper cutting of oil grooves in these boxes depends considerably the smooth running and freedom from heating in these journals that every engineer wishes to attain. As cup grease or hard oil is used almost exclusively on these bearings in place of liquid oil, these grooves are of far more importance than they were a few years ago. The grooves provide an easy passageway for the oil or grease as well as act as a chamber which keeps the journal constantly lubricated and thus reduces the danger of heating to a minimum.

Both halves of the crank pin and crosshead pin brasses should be grooved in the same manner. One very good way to have the brasses grooved is illustrated in Fig. 87. As the brasses are or should be relieved at the edges there will be two grooves already cut clear across the box. Starting at the lower left hand corner about three-eighths of an inch from the side of the box, gouge out a groove by cutting diagonally across the box to the upper right-hand corner, ending the groove about the same distance from the edge. The other groove is cut in the same manner by starting from the lower right-hand corner and ending at the upper left-hand corner. These grooves should be about one-quarter of an inch wide and one-eighth of an inch deep. The edges should be smoothed so that no rough projections will be left sticking up to cut the pin.

Babbitting Boxes.—This is not a particularly hard job, but the person doing it has to use his brains. The young engineer who lacks experience in doing this kind of work will not be able to do a satisfactory job in a short space of time but if he has perseverance and uses his brain he will be able to do it as well as his fellow-engineers who have babbitted boxes before.

Babbitting Main Shaft Boxes.—One of the heaviest jobs of babbitting on a traction engine is the main or crank shaft boxes. In order that the shaft may be reasonably certain to be placed in its proper position it may be desirable to babbitt both of them. A better chance is given to do a good job if the flywheel, clutch, main pinion, eccentric and connecting rod are detached from the crank engine shaft, making it much lighter. After removing the old babbitt clean the halves of the boxes thoroughly of all oil and grease by washing them in gasoline. It may be found necessary to rub the halves briskly with a stiff brush soaked in gasoline in order to remove all of the grease entirely. The shaft should also be cleaned in the same manner and then blocked in position.

A most particular and delicate job is to find the exact

REPAIRING OF ENGINES

position of the shaft and get it square with the engine frame. After taking off the cylinder head and removing the crosshead, pull the piston and rod out of the cylinder and remove the nut and gland from the stuffing box. Make a false head for the cylinder out of a piece of board so that it will exactly fit the studs on the cylinder. With a pair of dividers find the exact center of this head and bore a small gimlet hole through it. Now pass a piece of stout cord through the stuffing box, cylinder and small hole in

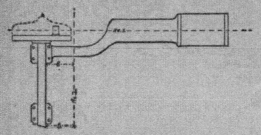

FIG. 88.—LINING MAIN SHAFT.

temporary head, tie this end to a stick or nail so that it is prevented from coming through the hole. Stretch this cord past the crank shaft and secure the other end to some convenient part of the engine. To get it in line measure the distance of the cord from the inside edge of the stuffing box. It must be the same from all sides. When the correct position of the cord has been located secure it in such a manner that it will not slip or become deranged. A plan of this method is illustrated in Fig. 88. Cord No. 2 should be stretched in the manner shown in the diagram. This cord should be exactly at right angles to cord No. 1. This method of lining a main shaft is equally adapted to center or side crank engines with single, double or compound cylinders. The shaft can either be blocked up at the ends or in the bearings. The former method is much the more desirable. When placing the shaft in position to babbitt the common style of split boxes, care should be taken to have the shaft an equal distance from the edges of the box. Now when the shaft is blocked up in this position with reference to the lower halves of the box, be sure that the distance B at the two points shown in the diagram is the same amount from the cord line No. . If it is found necessary to shift the shaft in order to m e the two measurements exactly the same, it can be done provided the shaft is kept in the same up and down posit by carefully measuring the distance of the shaft from the edges of the box. The shaft should be in its right posi on when these measurements are all made correctly as described. Before preparing to babbitt it, see that it is securely blocked so that it will not easily be moved. Ordinarily the crank disk comes up snugly against one end of one of the crank boxes.

To prevent the babbitt from running out of the box at the end, a piece of heavy cardboard, which is made to fit the shaft, is placed against the end of the box and also four or five liners which fit the shaft tightly the full length of the bottom side of the box. With soap or, better still, putty, carefully cement all crevices or holes at both ends and along the bottom of the shaft and box. Along the top of the box and shaft a trough is made out of putty or soap, a little higher than the top of the box, extending its full length and closed at both ends. To make it easier to put in the babbitt a cup is usually formed in the middle of the trough. It is a good plan to have both boxes ready so that both can be filled at the same time before the shaft need be bothered. If possible have the shaft and box heated as it is a hard matter to pour a good box if they are cold. Enough babbitt should be melted in a ladle or an old frying-pan to fill the box completely at one pouring. The babbitt

REPAIRING OF ENGINES

should be heated enough to char a small stick when placed in the melted metal. This will insure that it will run freely. The babbitt should be poured steadily until the box and trough are full. Having the latter filled allows a reservoir of metal to draw from as the box cools. Let the boxes stand long enough for the babbitt to harden before removing the cardboards and putty. In order to determine whether a good job has been done it will be necessary to slip the shaft out of the boxes. Sometimes when the box and shaft are not warmed or the babbitt not heated enough a poor box will be the result. If the box is not well filled or shows "ragged," i.e., the babbitt being in waves or rolls, it will have to be done over. The babbitt must be removed and the shaft relined. If the engineer has never done any babbitting he is apt to have a few failures at the start as this work requires a certain amount of skill and judgment. The best grade of babbitt should be used on crank shaft boxes as the cheaper grades are too soft for this kind of service.

The sharp edges of the box should be dressed by taking a light cut with a chisel. If the cardboards have been carefully fitted at the bottom all that is necessary is to remove the sharp edges of the box. The top edge where the babbitt has been poured must be dressed down level with the casting of the box and the edge slightly beveled like the other side. When both boxes have been trimmed the shaft should be put back in place and the top halves babbitted.

Before babbitting the upper halves place six or seven cardboard liners on each side of the shaft between the halves in order to allow plenty of room to take up future wear. These liners should rest tight against the shaft. Next replace the caps and bolt them down. If there is no hole in the cap through which the babbitt may be poured, it will have to be done from one of the ends. This requires a funnel to be made out of putty at each end. The ends are closed except for an air hole at its highest point around which is built the funnel. The air hole and funnel are made larger at one end than at the other. These funnels should be built up above the top of the box so that as the babbitt cools the metal in them will keep the box full. Both caps should be prepared to be rebabbitted at the same time and enough metal should be heated to do the work at one

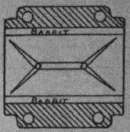

FIG. 89.—GROOVING MAIN SHAFT BOXES.

time without stopping. When the babbitt has had time to cool the caps should be removed and the edges dressed in the same manner as was done with the lower halves. The job of drilling oil holes in the babbitt may be saved by fitting plugs of wood in them which will extend down to the shaft. After the babbitting the oil holes may be opened up by pulling out the wooden plugs and reaming them smooth with a drill bit the size of the hole in the casting.

With a box which has two oil holes a heavy groove is cut from one oil hole to the other and somewhat lighter grooves from each hole towards the corners of the box. This method is illustrated in Fig. 89. These lateral grooves should terminate about three-eighths to one-half of an inch from the edge of the box, so as to prevent the oil from running out

REPAIRING OF ENGINES

of the bearing at these points. The center groove may be cut about one-quarter of an inch wide and almost as deep. The others should not be more in width and depth than three-sixteenths of an inch. Where there is only one oil hole the grooves may cross each other at the hole or they may be cut in the same as illustrated in Fig. 89. When using the latter method the center groove should be cut about one-quarter as long as the length of the box. The edges of all oil grooves should be rounded off enough to remove all sharp corners and rough places. The liners should be cut so that they will lack about three-eighths of an inch of touching the shaft, except at the ends where they should touch it. This will form an oil chamber on each side of the shaft. It will be found necessary to add one or two more liners to each side of the cap than was required when pouring. Be sure to put the same number on each side of the shaft. The bearings should be adjusted to a running fit. A close watch should be kept on them for the first few days of running until they have been worn smooth and during this period they should be well supplied with oil. As a new box is apt to wear down considerably at first, it may be necessary to take up a little on them after the first few days have passed.

If the boxes of the idler shaft, countershaft or axle are of the two-piece pattern they should be babbitted in exactly the same manner as the main shaft bearings explained above.

Babbitting Crank Shaft Boxes of the Quarter Box Type.—If the quarters are of brass, which have not become worn until new brasses are required, the old ones may be babbitted. Prepare the quarters for the babbitt by cutting three grooves in the quarters. One of these should be cut about three-quarters of an inch from each end and the other in the center. They should be about three-quarters of an inch wide and from one-quarter to three-eighths

of an inch deep. The edges should be left square.
is better not to smooth out the grooves as the babl
will stick much better to a rough surface. After c
ting these grooves, bore about three three-eighths-inch hc
in each groove at slightly different angles. These hol
which are for the purpose of holding the babbitt in pla
should not be drilled very deep. When all quarters h;
been prepared, place the bottom blocks in place *witht*
any liners under it and line up the main shaft as w
described on page 246. In order that the shaft may be at t
proper height in the box, it is necessary to have the cent
line of the shaft in line with the center line drawn throu
the cylinder. One way to determine that the shaft is
the same height in both boxes is to level the engine
placing a straight edge and level in the boxes *before* puttii
in the bottom blocks and then leveling the shaft after it
in place. Proceed in exactly the same manner as describ
for the split box type, to block the shaft, prepare and pot
the babbitt. After the bottom blocks have been babbitt
the shaft should be removed, the babbitt dressed level wi
the edges of the brass and the sharp corners remove
After this has been done the shaft should be replaced co
rectly in the bottom blocks in order that the adjustab
gibs on each side of the shaft may be babbitted.

To do this, first cut two shims from very thin cardboar
or heavy paper and place on each side of the shaft on th
bottom block. These shims must fit the shaft snugly a
they are for the purpose of preventing the babbitt of th
gibs from sticking to the bottom blocks. Put the gibs fo
both boxes in their proper position and unscrew the adjus
ing screws until the gibs are between one-quarter and three
eighths of an inch from the shaft. Close the ends wit
putty or soap in the usual way and then pour the babbit
Remove the gibs when time has elapsed for them to coc
and trim the edges, removing also any sharp corners. Th

REPAIRING OF ENGINES

...ge should be trimmed level with the brass. After
... have been placed back in their position, the top
... can next be babbitted. Five or six cardboard liners,
... must touch the shaft, should be placed on top of the
... The shims between the brass and the box caps should
... moved and the latter bolted down. The caps are pre-
... and babbitted in the same manner as those of the
... box type. When the babbitt has cooled, the caps
... be removed, the edges trimmed, oil grooves cut and
... oil holes reamed. The oil grooves cut in the top block
... be similar to that illustrated in Fig. 89. The boxes
... be adjusted to a loose running fit for a few days
... til they have become smooth. It is better to have them
... ook a little than that they should heat. After they have
... come smooth bearing they may be gradually taken up to
... proper place for smooth running.

Where the quarters of these boxes are cast, the old babbitt should be removed entirely, even to the small holes in
the blocks which hold the babbitt. Proceed to babbitt these
... boxes in the same manner as that described above.

Babbitting Crank and Crosshead Pin Brasses.—The
crank pins and brasses wear out faster than the crosshead
pin brasses and also cost considerably more than the latter.
Examine the crank pin after the brasses have been removed
and if it is found rough and scored, dress it (page 227).
The shims from both sides of the box must be removed, after
which the box should be replaced in the strap of the rod
and over the pin. To determine the amount of space which
will have to be babbitted in each half, a small wooden
wedge, one on each side, should be pushed between the pin
and the brasses. If this space should not be at least three-
sixteenths of an inch, the brasses must be hollowed out sufficiently to give this space. Even if the space is this
amount on each side it will be found that the top and bottom brasses are nearer together so that the brasses will

STEAM TRACTION ENGINEERING

have to be hollowed out to some extent in any case
detail cut, Fig. 90, shows how to hollow them out.
brasses are not cut out at the sides, a narrow strip
one-eighth of an inch wide being left. Several hol
drilled into the bottom of the brasses to hold the b
After having hollowed out the brasses and drilled the
they are ready to be babbitted. Set them in the strap
rod and place them as far apart as they will go. No

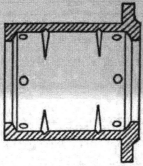

FIG. 90.—HOLLOWING OUT BRASSES FOR BABBITTING.

should be put behind the halves. Should an adju
wedge, block and screw be used to take up the wear it
be in place. Its position must be such that it will
the brasses to be apart as much as possible.

The rod must be blocked in such a position that the
pin will be exactly in the center of the brasses both
zontally and vertically. A small piece of wood mu
placed between the brasses at both top and bottom.
piece should fit tightly between the brasses and the pi
strap. These pieces will hold the brasses in positio
will divide them so that the babbitt will not be able t

ther, sticking the halves to each other. The cracks at back of the box must be sealed with either putty or, care being taken not to slip the brasses outward on pin. The outer edge of the pin and brasses should be a heavy cardboard washer fitted over them, which is held in place by a block of wood fastened securely against. The top of this washer must be cut off so that a space left in which to pour the babbitt, and it should have putty or soap placed around it tightly to seal the ends. A large funnel of putty to each half should be built at the top where the babbitt is to be poured in and where the washer has been cut to leave an opening. Care should be taken at all times to prevent squeezing any of the putty into the box or openings for the babbitt. Small air holes should be left at the top of each half and a small funnel built around them. Before pouring the babbitt the pin and brasses should be warmed. The best of babbitt should be used in this work, and it should be poured all at one melting until the funnels which lead to each half of the box are full of metal. By this method of preparation both halves can be babbitted at one time. After allowing the box sufficient time to cool enough for the metal to harden, the cardboard washer and putty will be ready to be removed and the brasses withdrawn from the strap. This will allow an examination of the babbitted part to be made in order to determine if a good job has been done. A good job is apt to be done the first time if the pin and brasses were sufficiently warmed and the metal well heated. They will have to be done over again if the halves are not entirely full or if they are very rough and uneven. When one half of the box has been properly babbitted while the job on the other half has been poor, it will only be necessary to rebabbitt the poor one. After both halves have been properly babbitted they should be removed and the edges trued up and relieved. Oil grooves should be cut according to the instructions given on page 245 and illustrated in Fig. 87. If washers were used on the old brasses to take up the side play, they should be removed before the box is babbitted, in case the babbitt will extend outside of the brasses to compensate for these washers. The new box will be the same length as the pin if the crank disk and the cardboard washer are properly puttied before pouring. After replacing the box in its strap, adjust the cap on the pin so as not to bind the box between it and the disk, and run the engine slowly for a while, closely watching the box to prevent it from getting hot and cutting. The box should be gradually adjusted to a running fit. The roughness of the bearing will cause it to heat very quickly if it is the least bit tight or not exceedingly well lubricated.

The babbitting of a crosshead pin box is simpler than the above if the Corliss type is used with a removable pin. The locomotive type of crosshead with solid pin is babbitted in the same manner as the crank pin brasses. To babbitt crosshead pin brasses, where the pin is removable, the connecting-rod should first be taken off and the crosshead pin removed. If found necessary, dress the pin down to a new bearing and hollow out the halves of the brasses. After these matters have been attended to, obtain a flat, smooth board and bore a hole in it the same size as the small end of the pin. This hole should fit the pin snugly and only allow it to go into the hole as far as the shoulder on the pin or its bearing surface. The brasses are now ready to be placed in the rod. Both the rod and the brasses should be laid on the board so that the pin may be placed in the hole in the board between the halves of the box. See to it that the pin is spaced equally distant from the sides of the halves and place small pieces of wood between the edges of the brasses. In order to obtain a box which is true it is necessary for the box to be level and the pin to stand plumb before the babbitt is poured. The

REPAIRING OF ENGINES

crevices around the brasses where they lie on the board are then sealed with putty and a ring of putty is made around the opening on top of the brasses. When pouring the babbitt fill the ring full so that the box will have a supply of metal to draw from as it cools. After it is cool the pin is removed and the sides and ends of the halves are trimmed down flush with the brass. The box also will be required to be relieved. The job will be completed when it has been replaced on the engine.

Babbitting a Crosshead.—To babbitt the locomotive type of crosshead, first remove it from the engine and with a sharp, narrow chisel cut a groove one-quarter to one-half inch wide by one-quarter inch deep in the center of the bearing surface nearly to each end. Next drill six holes, three on each side of the groove, three-eighths of an inch in diameter and about three-eighths of an inch deep in each bearing surface. These holes and the groove will effectively hold the new babbit in place. The crosshead is now ready to be replaced in position on the engine. In doing so, adjust the lower guide bars so that they will hold the crosshead in line with the travel of the piston or the center line of the engine. After three-sixteenths to one-quarter of an inch of liners have been placed under the upper guide bars (the quantity to be used at each end) the bars can be bolted down and a piece of wood placed between the bars and against the ends of the crosshead to prevent the babbitt from running out at the ends. Another piece of wood should be fitted against the side of the bars and the crosshead. This latter piece should have a groove cut at one end in which to pour the babbitt or it can be fitted with its upper edge just even with the edge of the crosshead and a putty trough made along the opening and slightly higher than it. The crevice at the ends should be closed with putty prior to pouring, but the crevice on top of the bar, between it and the crosshead, should be left open to permit the air to escape. The crosshead and bars should be heated quite warm just previous to pouring the babbitt. Always pour the trough full in order that there will be a supply to draw from as the babbitt cools. After the metal has had time to harden, the edges should be dressed off smooth and flush with the crosshead. In order to babbitt the bottom side of the crosshead simply turn it upside down and proceed in the same manner as above. This insures a much better job and is also a little easier to do than pouring them from below; besides, by this method the babbitt is much more apt to fill the holes and groove.

The Corliss type of crosshead is almost as easy to babbitt as the above type. After removing the shoes and cleaning them of all grease and oil, the old babbitt or the babbitt strips on their wearing surface should also be removed. If only babbitt strips have been used, three or four holes should be bored partly through the shoes at each end to help hold the new babbitt in place. If the gibs or shoes are of brass or the bearing surface is not babbitted either entirely or with strips, the shoes will have to be prepared for holding the metal. This is done by cutting three grooves in the face to each end, one of them being in the middle and one close to each end. These grooves should be cut the same width and depth as those mentioned in connection with the previous type of crosshead, and also the same size and number of holes should be bored. The shims between the shoes and the crosshead body must be removed, and if adjustment is made by sliding the shoes, they should be pushed back to allow as much space as possible between the shoe and the guide. If adjustment screws are used they must be backed out and the shoes drawn up against the crosshead body. When the shoes have been moved into this position they should be bolted so that they will not slip. Now place the crosshead nearly at one end of the guides and block it so that it is in line with the axes of the cylin-

REPAIRING OF ENGINES

der. Next close the ends between the guides and shoes with putty, taking care not to push the putty between the shoes and guide, in order that the metal can come out flush with the ends. A trough of putty should next be made along each side of the shoe. A slight space (about one-eighth of an inch) should be left between the putty and the shoe so that the metal can run freely. At the point where the babbitt is poured the trough must be enlarged into a funnel. When the metal has had time to harden, the babbitt should be cut away from the sides until it is flush

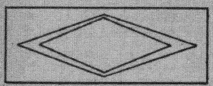

FIG. 91.—CROSSHEAD SHOE OIL GROOVES.

with the shoe. After this, first remove the shoe and with a scraping tool scrape the surface down smooth, then replace it and turn the crosshead upside down, thus bringing the other shoe to the lower guide. The above process should now be repeated for this shoe. When this shoe has been babbitted and trimmed on the sides, it must be removed in order to scrape the surface slightly. This scraping is to prevent having the crosshead too tight in the guides at first, as they cannot be loosened any other way until worn down to some extent.

Much better lubrication can be obtained if oil grooves are cut in the bearing surface of the crosshead shoes. One of the most successful ways of making these grooves is illustrated in Fig. 91. They are in the form of a diamond and should be made quite heavy at the points near the ends of the shoes and gradually taper down to light ones at the

FARM TRACTOR ENGINEERING

...should extend to within three-quarters of an inch and one-half of an inch of the side bearing... ...these should be adjusted in the same manner...

As is the case with all new bearings, it should... ...very closely for the first few days to see that... ...do not wear themselves and the guide.

Solid Boxes.—By a solid box is meant one in which... ...completely surrounds the... ...for the babbitt. As these... ...wear cannot be taken up without rebabbiting them.

To do this first remove the shaft and clean out all the old babbitt. This can be done by removing the box and placing it in the fire long enough for all of it to be melted and run out. This is by far the quickest and best way to get rid of the old metal, but in some cases it cannot be used and the metal must be cut out with a chisel. The grease will be got rid of in the burning out process and the boxes must be wiped clean before the shaft is replaced. Two layers of writing paper should be wrapped around the shaft at the point where it passes through the box. The ends may be glued down with anything that will make them stick, or they may be tied with light string or thread passed two or three times around the shaft. Anything will do to keep the paper in place around the shaft. The shaft and box should now be replaced and the former blocked so that it will be in the center of the box. Both ends should be closed with putty, with the exception of a small air hole on the top at one end and a large place at the other end. A small funnel should be made around the air hole and a large one at the other opening. A wooden plug should be placed in the oil hole until it touches the shaft, and the box should be thoroughly warmed before pouring in the babbitt. After allowing time for the box to cool, remove the shaft, plug and paper. This paper is placed on to prevent the metal

REPAIRING OF ENGINES

from sticking to the shaft and also to allow room enough so that the box will not be too tight. An oil groove should now be cut lengthwise of the journal across the oil holes to within one-half inch of each end of the box. Where the shaft is in an upright position it is only necessary to close up the lower end, insert the plug in the oil hole and put a little putty around them so that they will be tight. The babbitt will be poured from the top. Always wrap a couple of layers of paper around the shaft on any solid box or it may be so tight that the shaft cannot be moved.

Babbitting a Cannon Bearing.—Cannon bearings are largely used for both countershafts and rear axles. They are simply a large solid box and are no harder to babbitt than any other solid box. They are not babbitted for their whole length but only a portion of each end is done. The space between the babbitted bearings completely surrounds the shaft, but does not touch it, and acts as a reservoir for oil.

A detail plan of how to babbitt one of the bearings is illustrated in Fig. 92. The cut shows a countershaft, but an axle should be treated in exactly the same way. After withdrawing the shaft from the cannon bearing, remove all of the old babbitt from both ends and clean both of all grease and dirt. Now make two boards like A in the sketch. The hole in these boards must fit snugly, the shaft and the dotted circle must be the same diameter as the turned end of the bearing. The four blocks of wood must be spaced equally around the circle with their ends just flush with it. The center of the hole for the shaft *must* be exactly in the center of this circle. Wrap two or three thicknesses of paper around the shaft at the bearings after the shaft has been placed on end. With one of the boards on the lower end of the shaft, the blocks sticking up, slip the cannon bearing over the shaft and lower it until its end is between the blocks. Now slip the other board on the other end of the shaft with the blocks fitting over the edge of the cannon and the shaft will be in its proper position ready to have the bearings babbitted. Plug the oil hole with pieces of wood and place putty around the cannon

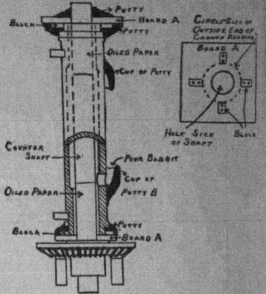

Fig. 92.—Babbitting Cannon Bearings.

where it rests on the boards, making a cup at the opening as shown in the illustration. After having babbitted one bearing turn the shaft end for end and babbitt the other box. After the bearings have had time to harden the shaft is withdrawn and oil grooves are cut in the top of each

REPAIRING OF ENGINES

bearing. These grooves should not be cut out to the outer end of the bearing but may be cut out at the inside end. Either axles or countershaft bearings may be very successfully babbitted in this manner without any danger of getting them out of line.

Babbitting an Eccentric Strap.—After laying the eccentric, which has previously been removed, on a smooth, flat board, bore three or four holes about three-eighths of an inch deep in each half of the strap. The inside of the strap should be cleaned of all grease and dirt. Next place the strap together in position on the eccentric with enough shims between the halves to allow about three-sixteenths of an inch space on each side of the eccentric. This space must be equally divided both at the top and bottom and endwise. The shims should fit against the eccentric so as to divide the babbitt. Now warm the eccentric and strap, put putty around the strap where it rests on the board, and pour the babbitt. A wooden plug should be inserted in the oil hole prior to pouring and reamed out when the babbitt is cool.

The following suggestions are given which are applicable in a general way to the babbitting of all types of boxes:

1. Where it is possible, have the box and shaft heated before pouring the babbitt.

2. Remove any old babbitt and thoroughly clean the box and shaft of all grease and dirt.

3. Be sure that the babbitt is hot enough to run freely while pouring.

4. Just previous to pouring the babbitt skim off the dross and dirt that collects while melting.

5. The boxes should always be "vented" so that the air can escape or a good box cannot be run.

6. Never run a box until certain that it is in its correct position for pouring.

Instead of putty, soap or clay mixed damp enough to stick

together may be used. Cylinder oil and flour mixed into a stiff dough may also be used instead of putty. This mixture can be used a number of times and is not open to the objection of melting, like soap, or of getting grit into the babbitt, like clay or mud.

Repairing Governors.—When the governor is not working properly from any cause whatever, the matter should be looked into at once and an effectual remedy applied. In repairing a governor, remember that it is a delicate piece of apparatus and on its accurate adjustments and easy working joints depends its accurate regulation of the engine.

The small driving gears are apt to become worn in course of time until the cogs begin to slip (Fig. 57). If some of the teeth have been entirely worn off, the gear will have to be replaced with a new one. But if they are only partly worn down, they may be repaired so that they will last a short time. In the latter case if upon examination of the gears it is found that the teeth do not properly mesh, take hold of the rods and pull them outward to see if the gear fastened to the revolving head raises more or less. If it does, the amount of this play should be taken up by loosening the small collar on the upright stem that holds the revolving head down and pushing it down against the head. Do not put it so tight or it will make the governor run hard. This will prevent the gear from rising away from the other. If this is not enough to make the gears mesh properly, remove the gear on the driving shaft and place a thin washer between the hub of the gear and the end of the driving shaft box. To not mesh the gears so deep that they will bind or else the governor will run hard and fail to work properly. The same precaution should be taken when repairing worn-out loose gears with new ones. The gear on the revolving head should be securely fastened. This gear is usually held in place by means of set screws which pass through the wheel into the revolving head.

REPAIRING OF ENGINES

The gear on the driving shaft is sometimes fastened by a screw, which is partly sunk into the shaft, or by a pin, which is driven through the hub of the wheel and the shaft. Very often, in cases where the pin is used, the hole through the shaft becomes worn until it is impossible to fasten the gear solidly. The best way to remedy this is to cut the pin off with a hack saw at the old hole and drill a new hole of the same size as the pin and the same distance from the end of the shaft as the old hole.

When the driving shaft has considerable play it is bound to get out of line and cause the governor to run hard. It also has a tendency to bind the gears and cause the belt to crawl off the driving pulley. To remedy this, remove the shaft and see if it requires to be renewed or if it can be repaired. If it is badly cut it is better to get a new one. If able to obtain a piece of steel shafting of the proper size, a new one may be made easily by drilling a hole for securing the gear in the same way as the old one, and cutting the shaft to the proper length, finishing the job by polishing it with a piece of fine emery cloth. If the old shaft is in good shape, simply clean it of grease and dirt.

The driving shaft bearing is very seldom babbitted, and when it is not, the box must be reamed out carefully with a large drill, but so as to allow room enough for the babbitt. If the bearings have been babbitted, simply remove the old babbitt and clean the box of all grease and dirt. After this has been done, wrap the shaft with a thickness of writing paper, place the shaft in position and block it in line with the center, or the revolving head. The shaft can be held in position by means of two or three small pieces of wood fitted in between the shaft and the box at each end. Place the shaft in an upright position, close up the bottom end around it, fit small plugs in the oil holes against the shaft, warm the box, and pour the babbitt in at the top. When
d, remove the shaft, ream out the oil holes, and e

small oil groove in the bearing. The job will be done upon replacing the shaft and adjusting the gears.

Fitting a New Governor Stem.—A new stem may be required after it has been used a long time, especially if the packing has been allowed to get dry and hard and it has become worn smaller at the point where it passes through the gland. In this case it cannot be packed tight enough to prevent its leakage without interfering with the perfect working of the governor. When the stem has become as badly worn as this, it will be necessary to have a new one in its place. If the stem comes in two pieces only the lower one will be required to be replaced, unless the upper one is found to be badly worn, when both should be replaced. If the stem is all in one piece the new one should also be in one piece in order to obtain the best results. In making a new stem procure a perfectly straight brass rod of the same size and length as the old one and provide it with the same number of holes located in the same position as in the old one. When drilling these holes in the stem or putting it in position be careful not to spring or bend it. A steel rod will also make a good stem, if smooth, but it is not nearly as good as a brass one. To straighten a bent stem lay it on a block of wood and use a wooden mallet. A short kink can seldom be taken out entirely and a new stem will be the best and quickest way out of the trouble. When a stem has been broken for any cause, do not attempt to fix it by riveting or welding it but replace it at once with a new one. A welded or riveted stem never proves satisfactory as it is apt to break again when least expected.

After long-continued use, especially where bad water is used and the engine primes quite a bit, the governor valve will become worn until it will leak sufficient steam to race the empty engine or when it is lightly loaded. As these valves cannot be removed, the only remedy is to replace them with a new one, but it sometimes happens that the seat will wear as much as the valve. Therefore replacing the valves will only correct half of the trouble. In such a case, to make a good job, there will be required both a new valve and its seat. Where the valve seat is removable the putting in of a new seat and valve is an easy matter. In putting in a new valve seat be careful not to spring it or chip its edges. The new valve may have to be rubbed down very slightly with fine emery cloth to make it work in the new seat freely without sticking, but do not do so unless absolutely necessary. In those types of governors where the valve seat is solid with the governor body, a new body will be required. As a new valve and body costs as much as a new governor it is much better to get the new governor in this case than to use the old one with a new body. Whenever a governor is taken apart for any reason, do not use a heavy gasket between the body and the top. Nearly all governors of recent construction have a ground joint at this point and therefore do not require a gasket. Should one be used, however, it should be as thin as possible. Badly worn posts or revolving heads can only be successfully repaired by using new posts and heads. This same principle may be applied as regards any spring that may become worn or weak.

Whenever an engine which is provided with a Waters governor does not give its full power and slows down considerably with a heavy load that it used to handle easily, it being reasonably certain that the valve is set properly and that the piston does not leak steam badly, the trouble will usually be found to be the spring inside the post that surrounds the revolving head. This spring is spiral and the governor stem passes through it. It is placed there for the purpose of raising the valve and is the only resistance that the stem gives against the balls. If, upon removing the arms that force the stem down, the stem drops to any

...AM TRACTION ENGINEERING

extent or does not spring clear up when pressed down, i[t is]
a certain indication that the spring has become weak o[r]
broken. Remove the same and if found broken replac[e]
with a new one, but if only weak it may be stretched u[ntil]
it is able to hold the stem up. This will make it last
quite a while but a new spring should be put in as soon
convenient. This only applies to the Waters governor.

When attaching a new governor to an engine, see t[hat]
the drive shaft is square with the main shaft and that [the]
pulleys are in line. This little matter has quite a bit to [do]
with the satisfactory operation of the governor. When [the]
top end of the governor stem has worn considerably
reason of the head continually holding the stem down, [use]
copper washers to take up the wear instead of iron on[es.]
Drive the governor with a thin, smooth belt nicely lac[ed]
not too tight. Small belt hooks are good things for t[he]
belt.

Repairing the Friction Clutch.—If the wood shoes th[at]
grip the flywheel require repairs, it is much better
get new ones from the factory, as they are inexpensi[ve]
and come all ready to put on. There is nothing difficu[lt]
or peculiar about the method of removing the o[ld]
shoes or in putting on the new ones, and, therefore, [no]
mention will be made of this part of the procedure. Whe[n]
in a hurry for new shoes they can be made, provided the[y]
can easily be obtained a piece of hard maple or dry oak [of]
the right size and thickness. The old shoes should not b[e]
used as an exact pattern in making the new ones. Th[e]
new shoes should be made thick enough to compensate f[or]
the future wear. To obtain the amount of thickness r[e]
quired, adjust the clutch so that the shoe holders are as f[ar]
away from the flywheel as the adjustment will permi[t,]
then throw the clutch lever in the position so that t[he]
clutch would be engaged if the shoes were in place. No[w]
measure the distance from the flywheel to the place whe[re]

REPAIRING OF ENGINES

oes rest in the shoe holders, and make the new shoes
m to these measurements. This will enable it to have
aximum amount of adjustment. Where the shoes are
d to fit the bevel of the flywheel they should be cut
form to this bevel or trouble will be had until the
have worn down to a fit. The shoes should be bolted
ly to their holders and after they have been used a
time they may be again drawn up.

ase of the breaking of any casting of the clutch, it is
lvisable to repair the break, but instead purchase and
a new piece. There is too much strain on this part
engine to risk any weak places in it. When the
g sleeve that engages or throws out the clutch becomes
the best thing that can be done is to buy a new sleeve.
her may be fitted in the groove to take up this wear
ere is danger of the split in this ring (it cannot be
solid) catching the fork arms and causing a general
. Sometimes, by careful management, the clutch ring
eeve may be babbitted, but the time that it will take
e cost of the babbitt are usually worth more than the
f a new ring and sleeve.

he main pinion and clutch spider are cast in one
and either part becomes broken, a complete piece will
to be purchased. Most builders at the present time
the pinion and spider in separate pieces, and the
, with the exception of the hole through which the
shaft passes, if not broken will last as long as the
. If this hole is kept well oiled at this point the
will be very slight; in fact, years of use will not cause
h wear to occasion any trouble, but should wear occur,
allow the clutch to wobble and this causes the shoes
h the rim of the flywheel at times when the clutch is
aged. This will induce the clutch to keep up a con-
jerking on the gearing all the time the engine is
g. These spiders are usually babbitted so that they

will not cut the shaft. In this case they may be rebabbitted in the same way as an ordinary solid box. The spider should be set exactly square with the shaft before pouring the babbitt or else the clutch will not work with any degree of satisfaction. If necessary the main pinion may be babbitted in the same manner. With reasonable care the box in the pinion will wear longer than the teeth on it.

If the turnbuckles become broken or worn out, it is best to get new ones from the factory, as very few blacksmiths are equipped to make them correctly. They are equipped with both right and left hand threads, which must be accurate if good results are to be obtained. On account of the tremendous strains which the clutch has to withstand it is not advisable to do any patch work on its parts, but instead replace the worn or damaged parts with new ones.

Fitting Cut-off Valves.—This is a job that an engineer should not attempt, if he can possibly avoid it, for the reason that it takes a long time and requires considerable skill and judgment to do a good job. The refitting of the valve to its seat is best done in a well-equipped machine shop. Do not try to do the work in the field unless the valve leaks so badly that the engine cannot be run with the valve as it is. At the shop the valve and seat are first planed off, then scraped to a steam-tight fit by hand. It is a pretty big undertaking for any man to do this job in the field where there is no power planer and the work has all got to be done by hand. To do this work, however, obtain a piece of very hard steel about three by four or five inches by one-eighth of an inch thick. This piece of steel must have perfectly straight and sharp edges. As the valve and valve seat are always cut about the same amount, both surfaces must be scraped down smooth and *perfectly flat*. This will be a slower process than scraping the hardest kind of wood with a piece of glass, but it can be done by a man who will have enough patience to keep at it until it is finished. Do

REPAIRING OF ENGINES

not use a chisel or file on the valve or its seat under any circumstances.

When putting in a new valve for any cause, such as a broken valve or holders, it is best to scrape the seat to a new bearing if it seems to be badly cut. If it is not cut considerably it is better to leave it alone. If it is desired to install a balance valve, the instructions which come with the particular make that is purchased should be followed. These valves are nearly always made to the measurements for the particular engine in which they are to be used, and very often have special instructions for the particular valve. When the valve and seat of an engine have become badly cut and therefore leak, one of the best ways to remedy the trouble is to install a balanced valve of some reliable make. A balanced valve is not so apt to leak as the plain valve, and requires less power to operate it, and by these means increases the power of the engine to a certain extent.

Fitting Piston Rings and Cylinders.—When piston rings become sufficiently worn to allow them to leak steam past the piston, considerable power is lost, besides a considerable waste of steam. This may be due to several different causes. The rings and the cylinder walls may be perfectly smooth and still be sufficiently worn to allow steam to blow past the piston. If it has had reasonable care and kept well oiled, this will not occur until after several years' use of the engine. The only remedy, in case the rings wear down so that they will not be able to expand enough to keep them in perfect contact with the cylinder walls, is to replace the old rings with new ones. Where an engine has had poor care and the cylinder allowed to run dry the rings will wear out very quickly. A cylinder in this condition will soon have its cylinder walls scored so that it will be impossible for the new rings to make a tight joint. The cylinder and rings have often been ruined in a

STEAM TRACTION ENGINEERING

time by not supplying the cylinder with oil
and thereby allowing the parts to run dry. Such
cylinders must be re-bored to make a tight joint again, and
this work cannot be successfully done outside of a well
equipped machine shop. If it is re-bored the piston should
also be fitted with a new head as the old one will be too
small for the increased size of the cylinder. New rings
should also be fitted in a case of this kind or the re-boring
will not be of any benefit. If the cylinder is not worn
and is in good shape, new rings will generally be all that
are required.

To test the piston rings to see whether they leak or not,
throw the reverse lever in position for the engine to run
over (i.e., the top of the flywheel away from the cylinder),
turn the engine to the crank end of the stroke with the
crank pin a few inches below the center, and remove the
cylinder head. With the engine in this position, block the
flywheel securely so that it will not revolve and turn on a
little steam. As the steam will be admitted to the crank end
of the cylinder it follows that should the rings leak the
steam can be seen coming out between the cylinder and the
piston. If the leakage amounts to very much it should be
stopped at once. Steam coming out of the steam port at
the head end of the cylinder indicates that the valve leaks
and should it be of any great extent the valve should also
be refitted at once.

Putting in new piston rings is a comparatively easy
matter. The piston must be removed from the cylinder and
the old rings taken off. Now try the new rings in the bore
of the cylinder to see if they will compress sufficiently to
enter the bore of the cylinder without the ends binding
together. If found too long they must be carefully filed
down until they will enter the cylinder without lapping.
Next try them in the grooves of the piston as they may be
a trifle too wide to work freely; if they are, dress them

REPAIRING OF ENGINES

carefully with a file until they do fit correctly. The should all be done on one side. To put them in the of the piston, first slightly warm the rings and carefully spread them until they will slip over the head. In order to prevent them from suddenly ping into the grooves use two or three wooden wedges. the ring into the groove as easily as possible by reing the wedges one at a time. Proceed in the same er with each of the rings. When putting the piston in the cylinder do not try to drive the rings past the terbore but compress them and hold them in that posi- by means of a piece of soft wire. The cylinder will quire an extra supply of oil for the first few days of nning in order to prevent cutting. When the rings leak ly and new ones cannot be procured for several days the ones may be spread and made to last for some time. To spread them place them on a hard wooden block and with a ball pein hammer strike very light blows all around the inside of the ring. This work must be done very carefully or a broken ring will be the result. Piston rings are made of cast iron and will break easily unless handled with care.

Packing an Engine.—Upon the proper performance of this duty depends a large share of the economy of the engine and also its smooth running qualities. Good packing should always be used, as a poor or inferior grade will soon cause the rods and glands to become badly worn, which will make them hard to be kept packed even with high-grade goods. Steam chest cover and cylinder head gaskets are usually very easy to make. A very good packing for these joints is an asbestos sheet packing interwoven with fine brass wire. While this packing is high in first cost, gaskets made from it will not burn or blow out and they can be used over again a number of times. Hard, thin cardboard makes a fairly good gasket for these joints if properly put on and not used with very high steam pressure. A gasket for all around uses, which is not expensive, can be made from either red or black sheet rubber. A high-grade spiral or ring packing is much the best type for piston and valve rods, while the packing to be preferred in small stems and valves is asbestos wick or candle wicking.

Fitting Gaskets.—A great deal depends on properly fitting a gasket if it is supposed to hold for any length of time. A gasket for a steam chest lid should be made out of one piece of gasket material. To make this gasket take a piece of gasket material of sufficient size and lay it on the face of the lid, holding it securely so that it cannot slip. With a hammer gently hammer the packing at the edge of the plate and for the bolt holes use the ball of a ball pein hammer. Now cut with a sharp knife the gasket along the lines shown by the hammering. Do this work of hammering and cutting carefully or else a poor gasket will be made, which will be apt to blow out in a comparatively short time. The gasket should be made so that it will project over the inside edge of the steam chest about one-quarter of an inch. They should be well oiled on both sides and the plate should be drawn up snugly at first. As soon as the steam chest becomes hot the bolts should again be tightened. If this matter is attended to strictly the gaskets will seldom blow out.

A gasket for a cylinder head should be cut and fitted in the same manner, only in this case the gasket must fit against the shoulder at the inside edge instead of lapping over the edge, as it does in the steam chest. Gaskets for governor flanges, heater ends, steam pipe flanges and throttle covers should be cut from the best of material and fit the flanges and bolts snugly. Oil both sides of a gasket and draw it up tightly when first put on, again tightening it as soon as it gets hot. Where two or more bolts or studs

REPAIRING OF ENGINES

are used to hold the joint together, do not tighten them one at a time, but instead give one a turn and then the next in order that the strain will be equal on all bolts. When nothing better is at hand cardboard may be used for these gaskets, but great care should be taken in cutting them out to see that they fit the parts nicely and do not become cracked or broken. They also are required to be well oiled. If a cardboard gasket ever starts to leak a new gasket is the only remedy for the trouble.

Packing Piston and Valve Rods.—The best grade of spiral packing should be used in these places whenever possible. The packing should be of the proper size to fit in the stuffing box without having to trim off the edges. To pack a piston or valve rod properly, remove the gland and take out the old packing from the stuffing box. Now take the packing and fit a ring of it around the rod so that the joint will lack about one-quarter or three-eighths of an inch of meeting. This is to provide for expansion which will cause the packing to buckle if no room is allowed between the ends. Cut the ends at an angle so that they will lap past each other to prevent leakage. Rings like the above should be continued to be cut until the stuffing box is full. In placing the rings in the box be careful that no two joints are in the same horizontal line. Screw the gland up with the fingers and never draw a stuffing box any tighter than is absolutely necessary to prevent leakage. If the gland or the nuts holding it work hard they should be oiled and worked back and forth until they can be screwed in easily. Throttle stems may be packed in the same manner as mentioned above.

Governor stems should be packed with asbestos or candle wicking well soaked in cylinder oil or graphite. First remove the cap and gland of the stuffing box and carefully take out all the old packing. After this wrap around the stem a strand of asbestos which has been previously soaked

STEAM TRACTION ENGINEERING

and push it down into the stuffing box. Continue doing this until the box is full, when the gland can be screwed up until it just prevents leakage. Another good packing for governor stems, or for that matter any small valve stem, is ordinary wrapping cord well soaked in cylinder oil. When packing the stems of steam or water valves, use asbestos or candle wicking soaked in oil, and wrap it around the valve stem the way the packing nut screws on, in order to prevent the nut from rolling the packing to one side. In packing the rod of a crosshead pump it is preferable to use either spiral piston or spiral hydraulic packing. The rings are cut and placed in the stuffing box in the same manner as in the stuffing box of a piston or valve rod. The old packing should always be removed before repacking any valve stem or stuffing box, as the old packing will become hard and flute or cut the rod, thus making it hard to keep it packed afterwards. Instead of spiral packing a second choice can be made from either asbestos or candle wicking or hemp. To pack a pump or piston rod with one of three packings mentioned above, proceed as follows: Remove all the old packing and take three strands of the packing and braid them tightly. They should be large enough so that the resulting braid will just fit the stuffing box. This braid should be well soaked in cylinder oil before being used in the stuffing box. When putting in a new gasket always scrape the surfaces perfectly clean of all old packing.

Where the hole in the gland has become worn considerably larger than the stem which it packs (as with crosshead pump, valve rods, throttle stems, valve spindles and water glasses) it is often very difficult to hold the packing in the stuffing box because it squeezes out between the gland and rod. This can often be remedied by taking a piece of soft wire (No. 10 to 14 for water glasses and valve spindles and heavier wire for larger glands) and form it into a ring

REPAIRING OF ENGINES

t will fit into the box or gland and over the spindle or
n, then wrap it tightly with a good hard wrapping cord
il it will fit the gland and rod snugly. Lastly, place the
g on top of the packing next to the gland and the pack-
will hold as good as a new gland.

APPENDIX

Boiler Arithmetic.—*Boiler Pressure.*—By this phrase is meant the pressure of the steam in pounds per square inch above atmospheric pressure. Many engineers do not realize that the total pressure in a large boiler is greater than that in a small one, nor do they understand why a large boiler with $\frac{1}{4}$ inch thickness of plate, having a tensile strength of 60,000 pounds, is not capable of withstanding the same pressure as a smaller boiler with the same thickness and strength of plate. In order that the reader may understand this statement, a problem involving the above will be worked out. Two boilers will be assumed, each 8 feet long and carrying 100 pounds of steam as recorded on the gauge. One of the boilers to be considered is 28 inches and the other 36 inches in diameter. The tubes may not be considered in this problem as the outside surface is all that is needed in order to figure.

The surface of the smaller boiler shell will be
$28 \times 3.1416 \times 96 = 8444.62$ square inches. The area of the two heads is $14 \times 14 \times 3.1416 \times 2 = 1231.51$ square inches. Adding these together and multiplying the sum by 100 gives 967,613 pounds, or 483.80 tons, as resisted by the smaller shell.

For the larger boiler the surface of the cylinder will be
$36 \times 3.1416 \times 96 = 10857.37$ square inches. The area of the two heads is $18 \times 18 \times 3.1416 \times 2 = 2035.76$ square inches. Therefore this boiler will have to withstand 1,289,313 pounds, which is equal to 644.66 tons. From this it will be seen that the larger boiler will require considerably heavier material than the smaller one to withstand the same pressure.

It can readily be seen from the above that there is a considerable amount of energy contained in a steam boiler. Is it to be wondered at that a boiler is a dangerous thing if not properly handled and that so much attention has been paid thus far to the

APPENDIX

avoidance of any carelessness and neglect on the part of the engineer?

A small boiler with a given grate area will give an engine as much power as can be obtained from a larger one. If the pressure can be maintained on the smaller boiler, it will cause the engine to give fully as much power as could be obtained by its several times its size. The size of the boiler has absolutely no influence on the pressure of the steam. Steam will be as powerful (volume for volume) when generated in a small boiler as from a larger one. If the above were not so and more power should be needed a larger engine would not be required or its pressure raised, but simply an increase in the size of the boiler. To illustrate this with an example, suppose that a large stationary plant has a boiler of two hundred (200) horsepower, having an engine with an 18 x 22-inch cylinder giving 125 horsepower and another with a 10 x 12-inch cylinder giving 30 horsepower. Now if the size of the boiler controlled the horsepower then the 10 x 12 engine would give as much power as the 18 x 22 one, but every intelligent engineer knows that it will not do it. Remember that 100 pounds of steam will push just as hard to the square inch if generated in a six-horsepower boiler as in the largest one ever built.

Another engineering term used largely by manufacturers and designers, which is not always understood, is that of "boiler pressure steam at the throttle." This means that the steam in the steam chest of the engine is at the same pressure as it is within the boiler. Such a condition is very seldom obtained, because the pressure is reduced more or less by condensation and friction in the steam pipe. The latter will vary according to the length of the steam pipe, its diameter, whether covered or not, the temperature of the steam and the surrounding atmosphere.

Steam Engine Arithmetic.—An engineer very often runs across little problems which are puzzling and which he may like to figure out to his own satisfaction. For this purpose a few problems will be worked out in order that he may grasp the method of procedure.

One of the most simple problems and one which bothers a

280

APPENDIX

good many engineers is the figuring the proper size of pulleys to give a certain speed or the correct speed of an engine to drive a machine at a certain speed with a certain sized pulley.

Ex. 1.—What should be the diameter of the threshing cylinder pulley to drive the cylinder 1,200 revolutions per minute (written r.p.m.) if the engine has a 40-inch flywheel and runs at 240 r.p.m.?

Rule.—Multiply the circumference of driver in inches by the number of revolutions it makes, and divide the product by the r.p.m. of the driven pulley; the quotient will be the circumference of the driven pulley in inches. Since the circumference is 3.1416 times the diameter (Rule 2, page 285) the diameter of the pulley is given by dividing the above quotient by 3.1416. Since also this factor (3.1416) appears both in the numerator and denominator of the calculation it can be omitted from both and the diameters used directly. Therefore in the above example we have

$$\frac{40 \times 240}{1200} = 8 \text{ inches as the diameter of cylinder pulley.}$$

Ex. 2.—At what speed should an engine having a 40-inch diameter flywheel run to drive a machine 600 r.p.m. that has a 14-inch pulley?

Answer
$$\frac{14 \times 600}{40} = 210 \text{ r.p.m. of driver.}$$

The unit of power is the "horsepower" (written H.P.) and is defined as the amount of power required to raise 33,000 pounds one foot in one minute, or one pound 33,000 feet in one minute. This may further be defined as foot-pounds, i.e., one H.P. is equal to 33,000 foot-pounds (written ft. lbs.). The total power of an engine is the total pressure in pounds per square inch (written lbs. per sq. in.) multiplied by the number of feet the piston travels in one minute and divided by 33,000. This gives the actual amount of work done in the *cylinder* and not the amount of H.P. *delivered* by the engine at the *flywheel*. The friction in the engine has to be deducted from the *cylinder horsepower* in order to arrive at the amount *delivered by the flywheel*. The amount of this friction will vary according to the condition and size of the engine and may be as much as ten per cent of

APPENDIX

the total cylinder power developed. The total pressure on the piston varies, being nearly the same as the boiler pressure from the beginning of the stroke to the point of cut-off and then rapidly dropping as the end of the stroke is reached, owing to the expansion of the steam behind the piston. This pressure can only be accurately measured by means of a steam engine indicator, but by using the Table of Expansions (page 178) an approximate answer can be obtained. In using this table the point of cut-off must be known. In using the Expansion Table of Average Pressures deduct 5 per cent for back pressure and clearance losses to obtain the mean effective pressure (written M.E.P.).

Ex. 3.—In order to work out engine problems, an engine with the following characteristics will be taken as an example:

Diameter of cylinder........................... 9 inches
Stroke of piston............................... 10 inches
Revolutions per minute (r.p.m.)................240
Steam pressure per sq. in......................148 pounds
Cut-off at .... ¾ stroke

Answer.—To find the area of the piston multiply the square of the diameter by the constant .7854. This is one-quarter of the factor 3.1416 used in the previous example. Taking the above piston the area will be $9 \times 9 \times .7854 = 63.6174$ sq. in. The steam table (page 178) gives the average pressure of 140 lbs. of steam at ¾ cut-off as 136 lbs., and this, minus the 5 per cent (or 40.8 lbs. back pressure), makes the M.E.P. 96 lbs. per sq. in. The travel of the piston expressed in feet will therefore be

$$\frac{10 \times 2 \times 240}{12} = 400 \text{ ft. per min.}$$

Having obtained the area and travel of the piston, also the M.E.P., the horsepower generated by the cylinder of the engine can now be easily obtained by putting the proper figures in the formula.

$$\frac{63.6174 \times 96 \times 400}{33000} = 74 \text{ H.P.}$$

To obtain the *brake horsepower* of the engine, which is the power *delivered by the flywheel*, it will be necessary to deduct the

APPENDIX

...per cent friction in the moving parts of the engine, thus

... × 90% = 66.6 Brake Horsepower (written B.H.P.).

...ble engines are figured exactly the same, first figuring the ... of one cylinder and multiplying it by two.

Pull on the Drive Belt.—Ex. 4.—What is the belt pull of ...ngine with a 40-inch diameter flywheel, a 10-inch stroke, a ... piston and a M.E.P. of 65 lbs. per sq. in.?

Answer.—The total average pressure on the pistons 63.6174 × ... = 4135.13 lbs.

...he travel of the piston in one revolution is

$$\frac{10 \times 2}{12} = 1\text{-}2/3 \text{ ft.}$$

...s pressure exerted in one revolution is 4135.13 × 1-2/3 = ...1.88. The circumference of the flywheel is $\frac{40 \times 3.1416}{12} =$

10.472 ft. Wherefore the actual pull on the belt will be

$$\frac{6891.88}{10.472} = 658.12 \times 90\% = 592.308 \text{ lbs.}$$

...rovided the belt does not slip nor is too tight.

Each inch of width with a single leather belt of standard grade will transmit one horsepower running at 600 to 800 ft. per min., 2 H.P. at 1,600 ft., and 3 H.P. at 2,400 ft. A 2-inch belt will give twice the power at the same speed. A heavy four-ply rubber or gandy belt will give 12½ per cent more power, a 5-ply belt 17 per cent, and a 6-ply or double leather belt about 23 per cent more power than a single belt.

Ex. 5.—What power will a single leather belt transmit that is 5 inches wide running over a 6-inch pulley at 1,200 r.p.m.?

Answer.—The circumference of pulley over which the belt travels is

$$\frac{6 \times 3.1416}{12} = 1.57 \text{ ft.}$$

The speed of the belt is 1.57 × 1200 = 1884 ft. per min.

Using the above rule it is seen that a belt running at this speed will transmit 2¼ H.P. per inch of width, therefore this belt will transmit 5 × 2.25 = 11.25 H.P.

APPENDIX

Ex. 6.—How much power will a 9-inch 6-ply drive belt t[ransmit] mit if the flywheel is 40 inches in diameter and runs a[t] r.p.m.?

Answer.—It has already been shown that the circumferenc[e of] the flywheel is 10.472 ft. The belt speed will be $240 \times 10.47[2 =]$ 2,513.28 ft. per min. According to the rule a 9-inch single [belt] running at this speed will transmit $9 \times 3.125 = 28.125$ [H.P.] Adding the 23 per cent for a 6-ply belt makes a total of 34[.6] H.P, that this belt will transmit without slipping if it is in g[ood] condition.

Capacity of Tanks.—To find the capacity of tanks divide [the] cubical contents in cubic inches by 231. This will give the [ca]pacity in gallons. To obtain the capacity in barrels divide [the] answer by 31.5 or multiply the cubic feet by 7.5. The ans[wer] in both cases will be in gallons.

Ex. 7.—What is the capacity of a square tank having [the] following dimensions—length 8 ft., 4 in., width 34 in., hei[ght] 29 in.?

Answer.—The 8 ft. 4 in. equals 100 in., therefore the cubi[c] contents of the tank will be $100 \times 34 \times 29 = 98,600$ cu. in., a[nd] the capacity

$$\frac{98,600}{231} = 426.84 \text{ gallons or } 13.5 \text{ barrels.}$$

Ex. 8.—What is the capacity of a round cylindrical tank [34] inches in diameter and 9 ft. (108 in.) long?

Answer.—The circumference of the tank is $34 \times 34 \times .7854 =$ 907.92 sq. in., and the cubical contents will be $907.92 \times 108 =$ 98,055.36 cu. in. The capacity will therefore be $\frac{98,055.36}{231} =$ 424.91 gallons. To obtain the number of barrels to fill this tan[k] we have

$$\frac{424.91}{31.5} = 13.49 \text{ barrels.}$$

Ex. 9.—What is the capacity of a tank of elliptical sha[pe] 36 in. wide, 28 in. deep and 90 in. long?

Answer.—The area of an ellipse is 3.1416 times the half su[m] of the major and minor axis, which in the example are 28 a[nd]

APPENDIX

respectively. This gives $3.1416 \times \dfrac{28+36}{2} = 100.53$ sq. in.

The cubical contents of this tank will, therefore, be $100.53 \times 90 = 9,047.70$ cu. in. The capacity, therefore, of the tank will be $\dfrac{9,047.70}{231} = 39.16$ gallons or 1.24 barrels.

Ex. 10.—What is the capacity of a tank 5 ft. in diameter at top, 8 ft. at the bottom and 11 ft. high?

Answer.—The mean diameter of the tank is
$$\dfrac{5+8}{2} = 6.5 \text{ ft.}$$

And the contents of the tank will be $6.5 \times 6.5 \times .7854 \times 11 = 365.014$ cu. ft. Since one cu. ft. contains 7.5 gallons, therefore the answer to the example will be $365.014 \times 7.5 = \dfrac{2,737.60}{31.5} = 8.91$ barrels.

While many other examples could be given, the above will be sufficient for most farm traction engineers. Before closing a few simple rules, which will come in handy in working out the above or other examples, will be given.

RULE 1.—To find the area of a circle multiply the square of the diameter by the constant .7854.

RULE 2.—To find the circumference of a circle multiply the diameter by 3.1416.

RULE 3.—To find the diameter of a circle with a given area, divide the area by .7854 and extract the square root of the quotient.

RULE 4.—To find the pressure in pounds per square inch of a column of water of a number of feet in height, multiply the height of water in feet by the constant .434.

The following constants will also be found useful:

The pressure of air is taken as 14.7 lbs. per sq. in. at sea level. This makes the total pressure per square foot of area $144 \times 14.7 = 2,116.8$.

One American gallon equals 231 cu. in. of water at 62° F. and has a weight of 8.34 lbs. There are, therefore, 7.5 gallons to the cubic foot of water, when the latter weighs 62.5 lbs.

APPENDIX

A cubic foot of fresh water weighs 62.4 lbs. at 62° F.
One square foot of half-inch boiler plate weighs approximately 20 lbs.
A cubic foot of iron weighs approximately 486 lbs.

EXAMINATION QUESTIONS AND ANSWERS

A large number of these questions are asked in the examinations for licenses to operate a steam traction or farm machine:

1. Q.—What is steam?
A.—Steam is an invisible gas which is generated from water by the application of heat.

2. Q.—What is water?
A.—Water is a compound of oxygen and hydrogen, 89 parts oxygen by weight and 11 parts hydrogen. It has its greatest density at 39° F., changing to steam at 212° F. and to ice at 32° F.

3. Q.—What is fire?
A.—Fire is the result of a rapid combustion or oxidation of solid organic matter or a gas so that both solid and gas glow and radiate heat.

4. Q.—What is flame?
A.—Flame is the luminous part of a current of hot gas when the rapid combustion or oxidation of the gas so takes place that particles of solid matter in the gas glow and give off heat.

5. Q.—What is coal composed of?
A.—Coal is composed of carbon and hydrogen mainly, with additional elements of sulphur, oxygen and earthy or mineral matters such as iron, lime, silica and the alkalies, the latter solid matter usually grouped together and called ash.

6. Q.—What is air composed of?
A.—It consists of 23 per cent of oxygen and 77 per cent of nitrogen.

7. Q.—What is heat?
A.—Heat is a form of energy which acts upon matter so as to be revealed to the senses of the body either as mechanical energy or by increase of temperature of the masses acted upon.

APPENDIX

8. Q.—What is the unit of measurement of heat?

A.—It is the amount of heat necessary to raise a pound of water from 62° F. to 63° F. and is called the British Thermal Unit (written B.T.U.).

9. Q.—What is latent heat?

A.—It is the heat in B.T.U.'s necessary to change a body from a solid to a liquid form or from a liquid to a gaseous state without effecting a change in the temperature during that change of state. Heat which effects a change in the temperature is called sensible heat.

10. Q.—What is the latent heat of water and of steam?

A.—The latent heat of water is 144 B.T.U. and of steam 966 B.T.U.

11. Q.—What is absolute pressure?

A.—It is the gauge pressure plus the atmospheric pressure of 14.7, both given in lbs. per sq. in.

12. Q.—What is the weight of a cubic foot of fresh water at 62° F.?

A.—62.5 lbs.

13. Q.—What is the weight of a cubic foot of cast iron?

A.—Approximately 486 lbs.

14. Q.—What is the weight of a cubic foot of ½-inch boiler plate?

A.—20 lbs.

15. Q.—What is the fusion point of iron?

A.—2912° F.

16. Q.—What is the fusion point of steel?

A.—2532° F.

17. Q.—What is the fusion point of lead?

A.—618° F.

18. Q.—What is a fire tube boiler?

A.—A boiler in which the products of combustion pass through the tubes or flues, the latter being surrounded by water.

19. Q.—What is a water tube boiler?

A.—A boiler in which the water is contained within the tubes, while the products of combustion circulate around them.

20. Q.—What types of boilers are used mostly in stationary work?

APPENDIX

A.—The return tubular and the water tube types contained in brick settings.

21. Q.—What kind of boilers are used in portable work?

A.—Boilers which are self-contained such as the locomotive and the fire flue types.

22. Q.—Do all boilers require settings?

A.—No. Self-contained boilers do not require settings.

23. Q.—What is meant by the phrase "boiler settings"?

A.—The foundation, means of supporting the boiler and the inclosing walls which also form the flues or passages for the hot gases.

24. Q.—What is the difference between the plain tube and the submerged tube upright boiler?

A.—In the plain tube or dry top boiler the upper ends are not covered with water, while in the submerged type they are covered with water all the time.

25. Q.—Which type of these boilers is preferable?

A.—The submerged type, for the tubes and tube sheet are less likely to be overheated and burnt.

26. Q.—What is the difference between a Cornish and a Lancashire boiler?

A.—A Cornish boiler has a single main flue running the entire length at the bottom while the Lancashire boiler has two flues of somewhat smaller diameter running the entire length of the boiler.

27. Q.—How are the Cornish and Lancashire boilers fired?

A.—They are internally fired.

28. Q.—How are vertical boilers fired?

A.—Internally fired except for a few cases.

29. Q.—How are return tubular boilers fired?

A.—Externally.

30. Q.—What are the principal advantages of a multi-tubular boiler?

A.—Cheapness and economical absorption of heat from the metal of the heating surface.

31. Q.—How are locomotive boilers fired?

A.—Internally.

32. Q.—What is the most common type of locomotive boiler?

APPENDIX

A.—The fire box boiler.

33. Q.—What are the principal disadvantages of this type of boiler?

A.—Imperfect circulation, difficulty of keeping water leg clean and the necessity of bracing and staying the extensive flat fire box surfaces.

34. Q.—What are its advantages?

A.—Economy of space and of fuel, quick steaming qualities and the ease with which it can be moved.

35. Q.—What are the advantages claimed for the internally fired boilers?

A.—They are self-contained and are more economical in the use of fuel than an external one.

36. Q.—What is meant by a self-contained boiler?

A.—A boiler that is independent of any masonry setting, cast iron fronts, tie rods, etc., thus permitting them to be easily moved from one place to another with very little trouble.

37. Q.—What are their other advantages?

A.—They are able to carry extremely high steam pressures besides being steady steamers. They are also very economical in the use of fuel.

38. Q.—Why are the tubes of a water tube boiler set at an angle?

A.—To help increase the circulation.

39. Q.—How much is this angle of inclination?

A.—Usually one inch in twelve.

40. Q.—Why are water tube boilers claimed to be safer than fire tube boilers?

A.—On account of the small amount of water contained in the tubes and also because of the smaller diameter of the tubes.

41. Q.—Name some of the advantages of a water tube boiler.

A.—Safety from explosion, tubes less likely to rupture and their quick steaming qualities.

42. Q.—What is a sectional boiler?

A.—A boiler made up of a number of sections or units. These sections are made of cast iron or steel and connected by narrow necks. The Harrison and Babcock and Wilcox boilers are the best known of this type.

APPENDIX

43. Q.—How is the grate area usually designated?
A.—In square feet.

44. Q.—How many square feet of heating surface is allowed per horse power in a steam boiler?
A.—It varies with the different types and makes of boilers, being as much as 15 sq. ft. per H.P., while in some cases it is as low as 8 sq. ft., the average being about 11 or 12 sq. ft.

45. Q.—How much heating and grate surface are required in a flue boiler?
A.—About 14 sq. ft. of heating and ¾ sq. ft. of grate surface per H.P.

46. Q.—The heating and grate surface of a tubular boiler?
A.—From 12 to 14 sq. ft. heating surface and ½ sq. ft. grate surface per H.P.

47. Q.—What is considered good evaporation?
A.—Seven (7) lbs. of water to one lb. of ordinary grade of coal.

48. Q.—How is boiler steel usually made?
A.—By the open hearth process.

49. Q.—What is the chief ingredient of all commercial iron and steel?
A.—Pure iron, of which they contain from 93% to 99%.

50. Q.—What is the principal difference between wrought iron and steel?
A.—The amount of carbon contained in them. To make steel from pig iron it is necessary to burn the carbon out of it and to make it from wrought iron it is necessary to add carbon to it.

51. Q.—Why is a steel boiler superior to an iron one?
A.—Because it is lighter and stronger.

52. Q.—Does a boiler plate become stronger as it is heated?
A.—Yes, to a certain extent. It becomes tougher and stronger until heated to about 550° F., beyond which point it gets weaker until at the fusion point it has no resistance at all. (See Q. No. 16.)

53. Q.—How would you compute the bursting strength of a cylindrical pressure vessel like a boiler?
A.—The diameter in inches multiplied by the pressure in pounds per sq. in. is the rupturing force. This is resisted by the

APPENDIX

area of plate on both sides in one inch of its length, multiplied by the tensile strength of steel expressed in pounds per sq. in. Expressed in symbols, this becomes:

$$PD = 2tf$$

If t is the thickness in inches and f the tensile resistance of the steel.

54. Q.—Is the head stronger than the longitudinal seams?

A.—Yes. The head or ring seams are twice as strong as the longitudinal seams. The head may bulge if not stayed.

55. Q.—What is the factor of safety of a boiler?

A.—It is the number of times that PD in the answer No. 53 will go into 2tf when f is taken as the ultimate strength of the steel reduced to allow for the weakness of the riveted joint and P is the working pressure. It is usually 6.

56. Q.—What is the efficiency of a riveted joint in a boiler?

A.—The efficiency of a joint is the ratio between its strength and that of the solid piece.

57. Q.—What is the number of handholes and where located on a traction or portable boiler of the locomotive type?

A.—Six—one in the rear head below the tubes, one in the front head at or about the line of the crown sheet, four in the lower part of the water leg; also where possible one near the throat sheet.

58. Q.—What is the number of handholes and where located in a vertical fire tube boiler?

A.—Seven—three in the shell at or about the line of the crown sheet, one in the shell at or about the line of the fusible plug when used, three in the shell at the lower part of the water leg. A submerged tube type shall have two or more handholes in the shell in line with the upper tube sheet.

59. Q.—What is meant by a boiler horse power?

A.—It is the evaporation of 34.5 lbs. of water at 212° F. into steam at 212° F. or the evaporation of 30.018 (Marks and Davis steam tables, 1909) pounds of water from feed water having a temperature of 100° F. into steam at 70 pounds pressure and is equal to the absorption of 33,479 B.T.U. per hour by the water in the boiler.

60. Q.—What is meant by the efficiency of a boiler?

APPENDIX

A.—It is the ratio between the heat units utilized in the production of steam and the heat units contained in the fuel used.

61. Q.—What are the standard rules for conducting boiler trials?

A.—The rules adopted by the American Society of Mechanical Engineers.

62. Q.—What is a safety valve?

A.—A valve attached to the boiler for the purpose of relieving excessive pressure.

63. Q.—Should each boiler in a battery have a separate safety valve?

A.—Yes, always, and no valve should be inserted between it and the boiler.

64. Q.—How many safety valves should a boiler have?

A.—At least 2, if not more, except in the case of a boiler requiring a 3-inch size or larger, in this case only one may be needed.

65. Q.—At what pressure should each of the safety valves be set?

A.—One valve shall discharge the steam at or below the maximum allowable working pressure. The remaining valves may be set within a range of 3 per cent above this maximum, but the range of setting of all of the valves shall not exceed 10 per cent of the highest pressure to which any valve is set.

66. Q.—How would you find the load to be placed on a safety valve?

A.—Find the area of the valve (its diameter in inches multiplied by itself and by 0.7854) and multiply this area in sq. ins. by the steam pressure in lbs. per sq. in. The spring must be able to balance this pressure.

67. Q.—How would you compute the weight of the ball to be put on the lever of a ball safety valve?

A.—By the principle of the lever. If the force computed in the last answer be called F and the distance to the fulcrum from the center of the valve spindle be called d, then it must be true that $Fd = Wb$. That is, the weight W multiplied by the distance from the fulcrum to the suspension point of such weight b must be equal to Fd.

APPENDIX

68. Q.—Into what two classes are furnaces divided?
A.—Into straight draft and down draft furnaces.

69. Q.—What is the object of a damper?
A.—It is to regulate the draft and so to keep the steam pressure constant.

70. Q.—What is a steam gauge?
A.—A gauge attached to the boiler to record the pressure within it in lbs. per sq. in. *above* the atmospheric pressure.

71. Q.—Is this *absolute* pressure?
A.—No, the absolute pressure is 14.7 lbs. in excess of that shown by the gauge.

72. Q.—What is meant by atmospheric pressure?
A.—It is the pressure of the air that exists on all objects on the surface of the earth. It is not felt by us as it is equalized, i.e., the pressure is exerted equally in all directions.

73. Q.—What is the purpose of a whistle which is attached to a boiler?
A.—For signaling.

74. Q.—What is a feed water heater?
A.—An apparatus for heating the feed water of a boiler before it enters the latter or comes in contact with the boiler heating surface.

75. Q.—What is the object of heating the feed water?
A.—To save fuel and also the boiler by decreasing the expansion and contraction.

76. Q.—What is a steam separator?
A.—A device designed to remove water, oil, dirt and other impurities from a current of steam. It is usually placed on the steam line between the boiler and the engine in close proximity to the throttle.

77. Q.—What is a steam trap?
A.—An appliance for removing the water of condensation from steam pipes, separators and similar apparatus without any waste of steam.

78. Q.—What is an injector or inspirator?
A.—An appliance for forcing the feed water into the boiler without the use of a pump or piston but by the use of a jet of steam.

APPENDIX

79. Q.—What is a water column?

A.—It is a tube with its two ends connected respectively to the steam and water spaces of a boiler and to which the gauge cocks and water glass are connected.

80. Q.—What is the object of a safety water column?

A.—To warn those in charge of a boiler that the water in the boiler is above or below its proper level.

81. Q.—What is a fusible plug?

A.—A plug made of brass having a centrally drilled hole which is filled with tin. The plug is threaded on its outer circumference so that it can be screwed into the boiler. The purpose of the plug is to give warning when the water gets too low in the boiler.

82. Q.—What are the principal requirements in the management of a boiler?

A.—(1) Proper firing; (2) a correct water level; (3) keeping it clean; (4) all appliances in good order, and (5) constant watchfulness.

83. Q.—Why should a boiler be always properly filled with water?

A.—In order to keep the sides of all surfaces of the boiler that are exposed to the flame and hot gases in constant contact with water. When iron or steel is heated more than 550° F. it gets weaker, therefore it is necessary to have the water in contact with the metal to prevent this high temperature from burning and weakening the sheets.

84. Q.—How often should a boiler be cleaned?

A.—It depends upon the amount of water evaporated as well as upon the quality and purity of the water being used in the boiler.

85. Q.—What is meant by priming?

A.—It is the water in the boiler being carried over into the steam pipe, eventually getting into the cylinder of the engine.

86. Q.—What are the common causes of priming?

A.—(1) Foaming; (2) too much water in the boiler; (3) boiler working beyond its capacity; (4) irregular firing; (5) sudden opening of stop or throttle valve.

87. Q.—What is meant by foaming?

A.—It is water carried over from the boiler the same as prim-

APPENDIX

ing but not in solid masses. The water does not lift in foaming as in priming but simply foams over, due to the dirt or grease in the water in the boiler.

88. Q.—What is the only effective preventive of foaming?

A.—The use of pure water.

89. Q.—What should be done in the case of low water?

A.—Cover the fire with ashes or green coal, leave the furnace door open and close the ash pit door.

90. Q.—Should the safety valve be opened, the engine stopped or the stop valve closed in such a case?

A.—No. Simply stop the source of heat.

91. Q.—Should the pump or injector be started or stopped?

A.—No. Leave them as they are, if they are not running. The sudden introduction of water upon the highly heated plates would cause too rapid an evaporation of the water. Do not start either until the boiler has had sufficient time to cool down.

92. Q.—How often should a boiler be inspected both externally and internally?

A.—At least once a year by a conscientious and competent boiler inspector.

93. Q.—What are the two principal ways of inspecting a boiler?

A.—The hammer and hydrostatic test.

94. Q.—How is the hydrostatic test applied?

A.—The boiler is filled full of water and a pressure which is 50% above the normal working pressure of the boiler is applied by means of a pump.

95. Q.—Why is it necessary that only the pressure required for safety be placed upon the boiler?

A.—In order to avoid straining the plates.

96. Q.—What is a check valve?

A.—A valve so constructed that a liquid or gas can only flow in one direction.

97. Q.—What is the best valve to use on the blow-off pipe?

A.—The Y valve or a plug cock packed with asbestos. These are far better than a globe valve.

98. Q.—How should a globe valve always close?

APPENDIX

A.—Against the pressure. The pressure should come against the under side of the disk and not on top of it.

99. Q.—What advantages has a gate valve over a globe valve?

A.—A fluid or gas can flow through a gate valve with less resistance than through a globe valve?

100. Q.—What is an angle valve?

A.—An angle valve is one in which the outlet above the valve is at a right angle with the opening below it.

101. Q.—What is a throttle valve?

A.—It is a valve which allows the weight or speed of the flow of steam from a boiler or source of pressure to be regulated and thus control the speed or work of the engine.

102. Q.—What is a relief valve?

A.—It is a valve so arranged that it opens outward when a dangerous pressure occurs.

103. Q.—What is meant by a double seat or double beat valve?

A.—A form of balanced valve having two disks and two seats so that the pressure shall come between them and make the pressure upward nearly balance the pressure downward.

104. Q.—Will one blow-off valve be satisfactory for all pressures?

A.—No. Two valves in tandem position will be needed above 125 lbs. pressure.

105. Q.—Of what material are the steam pipes in power plants made?

A.—Steel first, then wrought iron.

106. Q.—How is the size pipe designated?

A.—By its internal diameter.

107. Q.—Are boiler tubes designated in the same way?

A.—No. Their size is computed from the external diameter.

108. Q.—Why is this distinction made?

A.—Because boiler tubes are made more accurate as to size than an ordinary steel or wrought iron pipe and the outside surface of a tube is smoother than that of a pipe.

109. Q.—Is a bushing and a reducer used for the same purpose in pipe work?

A.—Yes.

110. Q.—How are steam pipes generally proportioned?

APPENDIX

A.—So as to permit a flow of 6,000 feet per minute in pipes carrying live steam and 4,000 feet per minute in exhaust pipes.

111. Q.—How much will the capacity of a pipe be increased by doubling its diameter?

A.—Four times; or generally as the square of the radius of the pipe.

112. Q.—What provision is made for the expansion and contraction of pipes?

A.—By expansion joints or bends in the pipes.

113. Q.—What is water hammer?

A.—When steam flowing in a pipe comes in contact with a quantity of water held in a pocket or recess of a pipe there will be an instantaneous condensation and lowering of pressure. The sudden rush of steam into this lowered pressure will sweep the water along at great speed and when such water fetches up against a bend or reduced section it makes a noise like the blows of a hammer.

114. Q.—Is water hammer considered dangerous?

A.—Yes. In direct proportion as the pressure of the steam and the quantity of water held in the pocket and the distance it can flow before it is checked.

115. Q.—How is the capacity in H.P. of a given sized pipe determined?

A.—Multiply the square of the diameter by 6. This relation holds if the speed of the engine piston is about the average of 750 feet per minute.

116. Q.—What is saturated steam?

A.—It is steam which carries all the water which it can carry as a gas at that temperature and pressure.

117. Q.—What is wet steam?

A.—Steam which carries more water per cubic foot than that due to its pressure and temperature. It will drop this excess of water on any surface which it strikes and go on as dry saturated steam.

118. Q.—What is the meaning of superheated steam?

A.—It is steam which carries in itself a higher temperature than that due alone to its pressure. It cannot be superheated in contact with water.

APPENDIX

119. Q.—What is an eccentric?

A.—It is a circular disk keyed to the shaft in such a manner that as the shaft revolves it will impart the same motion as a crank the length of which is equal to the radius of the eccentric.

120. Q.—What is the radius of the eccentric?

A.—It is the distance from the center of the shaft to the center of the eccentric sheave.

121. Q.—What is the travel of an eccentric?

A.—Twice the radius or throw.

122. Q.—Why is there an inside lap given to the plain slide valve?

A.—To cause the exhaust port to close earlier, thereby causing more or less compression in the cylinder according to the amount of the lap. It also keeps the exhaust from opening too soon, which would cause a loss of power.

123. Q.—Why is there outside lap?

A.—So that the steam may be cut off before the piston reaches the end of the stroke, thus causing the steam to work expansively.

124. Q.—What is meant by expansion?

A.—Using the full steam pressure on only a part of the stroke, then cutting it off so as to allow the expansive force of the steam to force the piston the remainder of the stroke.

125. Q.—What is meant by lead and what is its purpose?

A.—Lead is the amount the valve is opened when the engine is on dead center in order to "cushion" the piston. That is to absorb the shock of its sudden stoppage at the end of the stroke, thereby taking the strain of the stopping and reversal of the reciprocating parts off the crank pin.

126. Q.—What is a true dead center?

A.—When the crank pin, crosshead pin and center of the cylinder are in one straight line.

127. Q.—What is the angle of advance of an eccentric?

A.—Without outside lap or lead on the valve the eccentric would be set at an angle of 90° with the crank. Where the valve has both lap and lead the eccentric must be moved on the shaft sufficiently to move the valve the amount of the lap and lead. The angle between the eccentric and the 90° position is called the angle of advance.

APPENDIX

128. Q.—Does the piston travel the same distance while the crank pin makes the half revolution nearest the cylinder that it does while making the other half of the revolution?
A.—No.
129. Q.—In which half revolution does it travel the farthest?
A.—While making the half revolution nearest the cylinder.
130. Q.—What causes this?
A.—The angularity of the connecting rod.
131. Q.—What is a compound engine?
A.—An engine using steam in two or more cylinders, i.e., exhausting from the first into a second, third or fourth cylinder.
132. Q.—How are compound engines classified?
A.—By the number of cylinders. Two cylinders being called a single compound, three cylinders a triple expansion and a four cylinder quadruple expansion.
133. Q.—What is a non-condensing engine?
A.—It is an engine in which the steam after being expanded in the cylinder is discharged into the atmosphere or into some place where the pressure is above atmosphere as in the heating system or where the boiling point of water is above 212° F. or that corresponding to the pressure of the atmosphere.
134. Q.—What is a condensing engine?
A.—It is an engine in which the steam after having been expanded in the cylinder is discharged into a condenser where the temperature is below 212° F. because the pressure is kept below that of the atmosphere. The easiest way to condense the steam is by contact with water, either directly or in cooling pipes.
135. Q.—What is the object of so condensing the steam?
A.—It is to remove as much as possible the back pressure on the piston and thus increase the mean effective pressure on it throughout its stroke.
136. Q.—What is the first duty of an engineer before starting the engine in the morning?
A.—Ascertain if the water is at the proper level in the boiler
137. Q.—What is his last duty on leaving the engine for night?
A.—To see that there is plenty of water in the boiler and freezing weather that all the pipes are drained.

APPENDIX

138. Q.—How should an engine be started which has stood idle for some time?

A.—Slowly. This will allow the water of condensation time to escape and also allow the cylinder to become thoroughly heated.

139. Q.—What is an engine horse power?

A.—The amount of energy required to raise 33,000 pounds one foot in one minute or 33,000 foot-pounds.

140. Q.—What is the indicated horse power of an engine?

A.—It is the amount of work done by the steam in the cylinder as found by the use of an indicator.

141. Q.—What is the actual horse power?

A.—The amount of work delivered at the flywheel rim to the drive belt.

142. Q.—What is the rule for figuring the actual horse power of an engine?

A.—Multiply the area of the piston in square feet by the mean effective pressure by the speed of the piston in feet per minute and divide this product by 33,000. Deduct from this answer the amount of friction of the engine. (Assumed as approximately 10 per cent of the indicated H.P.)

143. Q.—What is the mean effective pressure in the cylinder?

A.—It is the average pressure during the entire length of the stroke of an engine which works expansively or has a cut-off action. When the boiler pressure is known by gauge, the cylinder or steam chest pressure should be 90 per cent of this. When the point of cut-off is known, the mean pressure can be ascertained from tables or ascertained by means of an indicator.

144. Q.—What is piston speed?

A.—The number of feet traveled by the piston in one minute. It is the stroke in feet multiplied by twice the number of revolutions.

145. Q.—What is the proper method of setting the valves on a duplex steam pump?

A.—Remove the cover on the steam chest, push piston to center of stroke. Both rocker arms must be vertical. Place both valves so as to cover the steam ports equally, adjust the valve blocks so that the space between the blocks and the valves will

APPENDIX

be the same on both sides. This play must be enough to allow the valves to slip sufficiently to open either port slightly without moving rocker arms. After these adjustments have been made push one valve against the block on valve stem so that one port is opened slightly. Now replace the steam chest cover and tighten it. The lead or opening spoken of above should not exceed 1/16 inch.

GLOSSARY

Certain technical words and phrases are defined in this section, which the reader may not find in dictionaries or elsewhere.

Actual horsepower is the amount of work which is delivered to the drive belt at the flywheel rim. It is always less than the *indicated horsepower* by the amount of friction in the moving parts of the engine and the back pressure in the cylinder.

Angle of advance of an eccentric is the angle between the center line of the crank and the center line of the eccentric. See page 194.

Angle valve is one in which the outlet above the valve is at a right angle with the opening below it.

Babbitt is a metal used for bearings, which is composed of alloys of lead and tin with sometimes a small amount of antimony, copper, or zinc added.

Back connection is that portion of a return flue boiler which is at the rear of the cylindrical flue and the tubes. See illustrations on page 4.

Baumé is a name given to an arbitrary scale on a hydrometer, which instrument is used to determine the density of liquids.

Bead is that portion of the end of a boiler tube which has been curved back so as to form a fillet thereby making a tight joint with the tube sheet.

Beader is a tool used to form a bead. See Fig. 25, page 68.

Blower is a small pipe, one end of which is attached to the steam space of a boiler and the other end of which points up in the center of the smokestack for a short distance. Its purpose is to improve the natural draft of the boiler if necessary. See page 16.

Blow-off pipe is attached to the boiler at its lowest and coolest point to allow boiler scale and sediment to be re-

GLOSSARY

moved. A "surface blow-off pipe" is one located at about the same height as the working level of the water in the boiler. Its purpose is to remove the scum and foam which forms on the surface of the water when working with certain kinds of water. See page 14.

Boiler filler plug is located on the top of the boiler and is used when it is necessary to fill the boiler by hand or to add a reagent to the boiler to prevent the formation of scale.

Boiler horsepower is the capacity to evaporate 34.5 lbs. of water at 212° F. into steam at 212° F.

Boiler priming plug. Same as boiler filler plug.

Bore is the diameter of the engine cylinder. See illustration 36, page 104.

Box is the name given that part of the engine which carries a revolving element.

Brass-bound is the name given to the condition when the brasses of a pin bearing bind the pin due to unequal wear. See page 227.

Bridge is that part of the seat of a slide valve of an engine which separates the steam from the exhaust passages. See Fig. 36, page 104.

Bull pinion is a gear of small diameter and large face suitable for heavy driving.

Bushing is a short cylindrical piece of tubing placed around a pipe or shaft to make it fit snugly in its support and capable of easy replacement in case of wear. It can also be a short piece of stock threaded on the inside and part way on the outside. As such it is used to reduce one size of pipe in line to another.

Cannon bearing is one in which the length is considerably greater than its diameter. See page 261.

Calking is to upset the edge of the metal at a joint so that the latter becomes water and steam tight.

Check valve is one that is so constructed that a liquid or gas can only flow in one direction.

Clearance is that space at each end of the engine cylinder between the end of the stroke and the cylinder heads. Its volume includes those of the admission ports in addition.

Compensating gear is that mechanism which allows a machine to take a corner. It does this by allowing one of the driving wheels to roll through a different distance from the

GLOSSARY

other while both are driven by power. See page 146.

Clinker is the irregular mass of incombustible partly melted matter left by coal in burning.

Connecting rod is a rod connecting the reciprocating piece (crosshead) of a steam engine with the crank.

Counterbore is that space at each end of the engine cylinder which has a larger bore than that of the working portion of the cylinder. See Fig. 36, page 104.

Countershaft is one that is placed intermediate between the driving or main shaft and the driven member.

Crank is the device on a steam engine which changes the reciprocating motion to rotary. It is attached at one end to the connecting rod and at the other end to the main driving shaft.

Crosshead is a block sliding between two guides, which makes the piston rod of a steam engine move in a straight line.

Crown sheet is that portion of a locomotive boiler directly over the fire.

Cut-off is the point in the stroke of the piston at which the steam entering the cylinder from the boiler is shut off by means of the closing of the steam valve.

Damper is a valve placed in the smokestack or uptake to regulate the draft so as to control the fire.

Dead center is the position of the engine when the crank pin, crosshead pin and the center of the cylinder are in one straight line. It occurs at each end of the stroke.

Differential is another name for compensating gear.

Drain cock is placed at the bottom of the engine cylinder, one at each end, for the purpose of draining off any water that may collect at the bottom of the counterbore.

Draw filing is to cause the teeth of a file to cut along their length instead of crosswise.

Dress is to remove the rough places of a bearing so as to leave a smooth finish, by scraping, filing, or otherwise.

Eccentric is that portion a steam engine which n the steam valve. See Fi page 117.

Efficiency of a riveted is the ratio which the st of the riveted joint be the strength of the plate.

Ell is a pipe fitting is

GLOSSARY

the two ends are at 90° with each other.

Engine horsepower is the amount of energy required to raise 33,000 lbs. one foot in one minute, or 33,000 ft. lbs.

Expansion is allowing the expansive force of steam to do work in forcing the piston in the cylinder a certain portion of the stroke with a lowering of its pressure in the process.

Factor of safety is the ratio in which the load which is just sufficient to overcome instantly the strength of a piece of material is greater than the greatest safe ordinary working load.

Feed refers to that part of the boiler operation which has to do with keeping the boiler supplied with water. Derivatives of this are feed water, feed-water heater, feed pipe, and feed pump.

Ferrule is a bushing or thimble inserted in the end of a boiler tube plate to spread it and make a tight joint. See bushing.

Fire box is that portion of an internal fired boiler in which the fire is built. See Fig. 1, page 2.

Fire test of oil is the temperature at which it will maintain a continuous flame if allowed to burn.

Flue sheet is that portion of a boiler fire box in which the tubes are located.

Follower is the lighter portion of a two-piece piston which is bolted to the larger portion which contains the piston rings.

Follower plate is the disk that holds the packing rings of the piston head.

Friction clutch is the device designed to permit quick application or disconnection of power.

Fusible plug is a brass plug with a tin center screwed into the boiler at the lowest permissible water level to give warning when the water gets too low in the boiler. See Fig. 12, page 15.

Gasket is a thin flat annular packing of rubber, leather, or metal sheet placed between two flat metallic surfaces which are bolted together to make the joint steam or water tight.

Gate valve is one in which a wedge-shaped disk is moved across the openings or orifices to open or close them.

Gauge cock is a conical plug valve screwed into the boiler to indicate the level of the

GLOSSARY

water contained therein. See page 9.

Gauge pressure is the amount in lbs. per square inch above atmospheric pressure as indicated on the steam gauge.

Gland is that part of the stuffing box which compresses the packing.

Globe valve is one which has a spherical body in which a disk is raised by a screwed spindle from a cylindrical or conical opening in a transverse partition or seat.

Governor is the device for regulating the speed of an engine under varying conditions of load and pressure.

Hack saw is a close toothed metal cutting saw.

Heating surface is the amount of surface in a boiler surrounded by water on one side and by flame or heated gases on the other.

Heel bar is a bent bar of particular shape which is used in the process of riveting a patch on a boiler. See page 85, Fig. 31.

"Hooking up" is an expression used when changing the position of the link block in a Stephenson link reverse gear.

Idler shaft is one used in transmitting power or motion but from which no useful work is taken off.

Indicated horsepower is the amount of work done by the steam in the cylinder as determined by the use of a steam engine indicator.

Injector is an appliance for forcing the feed water into the boiler by means of a jet of steam at boiler pressure. See Fig. 18, page 32.

Jam nut is an extra nut used to secure the principal nut from working loose.

Journal is that part of the shaft or axle which rotates in a bearing.

Lap is the amount which the valve projects over the port openings at each end when such valve stands in its central position. See Fig. 67, page 174.

Lead is the space or opening between the outside edge of the valve and the edge of the port on the steam end of the cylinder when the engine is on its dead center. See Fig. 67, page 175.

Lifter rod is the one which supports the link of a Stephenson link reverse gear. See Fig. 81, page 199.

Liner is a small thin piece of metal placed between two flat surfaces for the purpose

GLOSSARY

of alignment. See also shim.

Mean effective pressure is the average pressure during the entire length of the stroke of an engine which works expansively or has a cut off action.

Nipple is a short piece of pipe threaded on each end.

"Over" is a term used to express the direction in which an engine travels. An engine is said to run "over" when the crank pin travels away from the cylinder in its top position.

Packing is either a metallic or fibrous material used to prevent leakage of steam or water from a confined space into the outer air. It is made in various forms and sizes.

Packing nut is the cap to a stuffing box to force the gland against the packing.

Pet cock is any small plug cock opening into the air from a confined space through which pressure can be relieved, water drained, or satisfactory working determined. Drain cocks are usually pet cocks.

Pillow block is another name for a large or important journal bearing.

Pin is a short cylindrical or taper piece of steel which joins two pieces of mechanism together, such as the crosshead and connecting rod (crosshead pin).

Pinch in the crosshead and crank pin brasses is caused by the faster wear of the brasses in line with the push and pull of the connecting rod than at the top or bottom. Same as brass-bound. See page 227.

Pin clutch is another design to permit quick application or disconnection of power. See friction clutch. See Fig. 47, page 120.

Piston is the circular disk in the cylinder against which the steam operates or which displaces the water. See Fig. 36, page 104.

Piston ring is a split circular ring placed in a groove around the piston to make it fluid tight in its chamber. See Fig. 36, page 104.

Ply of a belt is a single thickness of the material. A two-ply leather belt is one having two single pieces of leather cemented together, one on top of the other, so as to have a thickness twice that of a single belt.

Priming is the passage of water from the boiler through the steam pipe, thus reaching the cylinder of the engine. See page 53.

GLOSSARY

Radius link is the curved slotted element in a reversing valve gear. The radius of the center of the slot is the distance from such center to the center of the engine shaft.

Reach rod is that portion of a reverse valve mechanism which connects the reverse lever to the element of a valve gear by which the valve is made to change from forward to reverse motion. See Figs. 82, 86, pages 204, 220.

Reamer is a tool for enlarging, truing, and sizing holes after they have been made in a machine.

Relief valve is one so arranged that it opens outward when a dangerous pressure occurs.

Relieve is to cut away the material. Used in the book in the sense of cutting away part of the brasses in a bearing so as to prevent the shaft or pin from becoming brassbound.

Ripper or ripping chisel is a tool used in boiler repair work. See Fig. 25, page 69.

Revolutions per minute. The speed of an engine is given in terms of this unit. It is the number of complete turns or revolutions of the crank in a minute. Another way to state it is the number of double strokes of the piston in a minute.

Scale is the deposit on the tubes and other surfaces of the boiler, of the soluble mineral salts contained in the boiler water and precipitated during the process of the change from water to steam. See page 54. Scale will also contain mud or sediment brought into the boiler by the feed water.

Seam is the point of contact of two edges of a boiler plate.

Seize is a term used when a box or bearing, by reason of heat or deformation, grips or pinches the journal which ought to turn freely within it, thus increasing friction or stopping motion. The pinching may be sidewise or endwise.

Set collar is one containing a set screw, used to prevent endwise motion of a shaft or pulley which turns freely on it.

Set screw is one with a pointed end which is screwed radially through a small hub to prevent it from turning on its shaft.

Shim is the same as a liner.

Skid engine is one in which the boiler and engine are

GLOSSARY

mounted on two parallel pieces of timber which are bolted together to form an underframe, so that it can easily be moved on rollers, to distribute the weight when the ground is soft.

Slab is a piece cut from the side of a round log in order to square the sides.

Sleeve is a hollow cylinder sliding upon a shaft. Its length is greater than the diameter of the shaft to which it fits.

Snap ring is another name for an elastic piston or metallic packing ring.

Staybolt is one used in certain parts of a boiler to brace the portions which are parallel, or nearly so, against bulging due to the internal pressure.

Steam chest is that part of the engine in which the valve operates. It is always located next to or above the steam cylinder.

Steam drum is a dome or cylindrical drum riveted on top of the boiler. The steam pipe to the engine is connected to the upper portion of this drum.

Stop valve is the same as a gate valve.

Stroke is the maximum distance which the piston travels in the cylinder while the crank turns through an arc of 180°.

Stud is a threaded bolt, one end of which is screwed into a tapped hole in the engine. The other end is threaded to receive a nut.

Stuffing box is that portion of the steam chest or cylinder through which the rod moves back and forth. It consists of a cylindrical box or cavity in which the packing is placed, which latter is compressed by a gland. A packing nut or bolts hold the gland securely in place.

Tap is a tool used to make a screw thread in a hole.

Tap screw is a short bolt used in a tapped hole.

Tender is that part of a tractor on which the supply of fuel and water is carried. It may be built in with the tractor or it may be a separate unit which is attached to the machine by chains.

Tensile strength is the amount of pull expressed in pounds per square inch which a bar of metal will withstand before it breaks.

Thimble is another name for a bushing, ferrule, or sleeve.

Throttle is the valve which controls the amount of steam

GLOSSARY

entering the valve chest from the boiler.

Throw of an eccentric is the distance from the center of the shaft to the center of the eccentric sheave.

Traction engine is a heavy moving steam engine used for dragging heavy loads on common roads or across fields.

Tram is a piece of steel of a definite length ground to a sharp point at both ends and used for measuring. One or both the ends may be bent at right angles.

"Under." An engine is said to run "under" when the crank pin travels away from the cylinder in its bottom position.

Valve cap is the cover or plug which is screwed into the top of a check valve.

Viscosity of a liquid is its resistance to its tendency to flow. It is given in terms of the number of drops per minute which will pass through a glass tube of known cross section. The liquid must be at a prescribed temperature.

Wash-out hole is one placed at the lowest part of the boiler for the removal of scale and sediment. A boiler is provided with a number of these holes placed in different parts.

Washer is a thin cylindrical piece of metal cut out round in the center which is placed over a bolt next to the nut. Several of them may be used under the nut when the thread of the bolt is not long enough.

Water column is a tube with its ends connected respectively to the steam and water spaces of a boiler so that the gauge cocks and the water glass can be conveniently connected.

Water leg is the side and front end of the fire box of a locomotive boiler and the circular portion around the fire box of a vertical boiler filled with the water to be evaporated.

INDEX

Adjusting, crank pin bearing, 225
  crosshead, 230
  crosshead pin bearing, 227
  eccentric, 233
  main bearing, 228
  steam gauge, 95
Alignment of gears, 157
Allen valve, 181
Angle, of advance, 194
  valve, grinding, 91
Arithmetic, of boiler, 278
  of engine, 280
Arnold reverse gear, 215
  setting of, 216
Average pressures, table of, 178
Axles, front, 161
  rear, 160

Babbitting crank shaft boxes, 251
  cannon bearing, 261
  crank pin brasses, 253
  crosshead, 258
  crosshead pin brasses, 253
  eccentric strap, 263
  main shaft boxes, 246
  solid boxes, 260
Baker balanced valve, 192
Bearing, cannon, 261
  main, 145
Bearings, adjusting crank pin, 225
  adjusting crosshead pin, 227
  adjusting main, 228
Blow-off pipe, 14

Blow-off valve, 14
Blower, 16
Boiler, low water in, 62
  patching a, 79
  preparing, for storage, 66
  priming, 52
  return flue, 3
  starting new, 42
  testing, 99
  vertical, 2
Boiler accessories: blower, 16
  blow-off pipe, 14
  blow-off valve, 14
  fusible plug, 15
  gauge cocks, 9
  glass water gauge, 8
  injector, 32
  safety valve, 12
  steam gauge, 10
Boiler arithmetic, 278
Boiler cleaning, 50
Boiler explosions, 64
Boiler feed pipe, 38
Boiler feed pipe fittings, 38
Boiler feed pumps: Clark, 24
  crosshead, 18
  geared, 22
  Marsh, 25
Boiler foaming, 52
Boiler hand-hole, 87
Boiler locomotive, 6
Boiler repairing tools,
Boiler scale, 54
Boiler tubes, 49

INDEX

Boilers, classification of, 2
Box type engine frame, 106
Boxes, babbitting crank shaft, 251
    babbitting main shaft, 246
    babbitting solid, 260
    hot, 150
Brasses, babbitting crank pin, 253
    babbitting, crosshead pin, 253
    oil grooves in crank pin, 245
      in crosshead pin, 245
    fitting crank pin and, 242
    fitting crosshead pin and, 238
Butterfly throttle, 122

Cannon bearing, babbitting, 261
Centering an engine, 194
Check valve, grinding, 89
Clark boiler feed pump, 24
Classification, of boilers, 2
    of engines, 103
Cleaning boiler, 50
Clutch friction, 118
    repairing, 268
Clutch pin, 120
Coal firing, 44
Cocks, cylinder, 124
    gauge, 9
    stop, grinding, 93
Compensating gear, 146
Compound engine valves, 188
Compression point, 175
Connecting rod, 111
Corliss crosshead, 109
Countershaft, 159
Crank pin bearing, adjusting, 225
Crank pin brasses, fitting, 242
    babbitting, 253
Crank pin oil grooves in brasses, 245

Crank shaft, 113
Crank shaft boxes, babbitting, 251
Crosshead, adjusting, 230
    babbitting, 257
    Corliss type, 109
Crosshead feed pump, 18
Crosshead fitting, 234
Crosshead locomotive type, 108
Crosshead pin bearing, adjusting, 227
Crosshead pin brasses, babbitting, 253
    oil grooves in, 245
Crosshead pins, fitting, 238
Cups, grease, 135
    oil, 132
Cut-off valve, fitting, 276
Cylinder, 104
Cylinder cocks, 124
Cylinder fitting, 271

D slide valve, plain, 174
Direct valves, 178
Displacement lubricator, 124
Double connection lubricator, 127
Double eccentric valve gear, 198
    setting of, 200
Double-ported balanced valve, 182
Double-ported piston valve, 179

Eccentric, 117
    adjusting, 233
    plain, valve gear, 193
      setting of, 194
    table of position of, 180
Eccentric strap, babbitting, 263
Engine, arithmetic of, 280
    centering an, 194
    knocks in, 155
    packing an, 273
    starting, 149
    storage of, 168

INDEX

Engine frame, box type, 106
  girder type, 108
Engine mountings, 142
Engines, classification of, 103
Examination questions, 286 to 302
Expansion, 177
Explosions, boiler, 64

Feed pipes, 38
  fittings, 38
Firing, with coal, 44
  with straw, 46
  with wood, 45
Fitting crank pin, 242
  crosshead, 234
  crosshead pin, 238
  crosshead pin brasses, 238
  cut-off valve, 270
  cylinder, 271
  flues, 70
  gaskets, 274
  gate valve, 94
  gauge glass, 88
  piston ring, 271
  slide, 234
  throttle, 94
  tubes, 70
Flues, fitting new, 70
Flywheel, 118
Foaming, boiler, 52
Friction clutch, 118
  repairing, 268
Front axle, 161
Fusible plug, 15

Gaskets, fitting, 274
Gate valve, fitting, 94
Gauge, glass water, 8
  fitting, 88
  steam, 10
    adjusting, 95
    testing, 95

Gauge cocks, 9
Gear, alignment of, 157
  compensating, 146
  lubrication of, 152
Geared boiler feed pump, 22
Gearing, traction, 145
Giddings valve, 183
Girder type engine frame, 108
Glass water gauge, 8
Globe valve, grinding, 91
Gould balanced valve, 188
Governor, 136
  repairing, 264
  stem repairing, 265
Grates, 48
Grease cups, 135
Grinding leaky angle valve, 91
  leaky check valve, 89
  leaky globe valve, 91
  safety valve, 94
  stop cock, 93

Hand-hole, boiler, 87
Handling a traction engine, 169
Hot boxes, 150
Hydrostatic lubricator, 126

Idler shaft, 160
Indirect valves, 178
Injector, 32

Knocks in engine, 155

Lap, 174
Lead, 175
Leaks, staybolt, repairing, 76
  rivet, repairing, 86
Leaky check valve, grinding, 89
Leaky seams, repairing, 78
Link reverse, 198
  setting of, 200
Locomotive boiler, 6

INDEX

Locomotive throttle, 123
Locomotive type crosshead, 108
Low water, 62
Lubrication, 152
  of gears, 152
Lubricator, displacement, 124
  double connection, 127
  hydrostatic, 126
  single connection, 128

Main bearing, 115
  adjusting, 228
Main shaft boxes babbitting, 246
Marsh boiler feed pump, 25
Marsh reverse gear, 210
  setting of, 211
Mountings, engine, 142

Newton balanced valve, 190
Non-leak balanced valve, 191

Oil cups, 132
Oil grooves in crank pin brasses, 245
  in crosshead pin brasses, 245
Oil pumps, 131

Packing engines, 273
  piston, 275
  valve rods, 275
Patch, boiler, 70
Peerless reverse valve gear, 223
Pin clutch, 120
Piston, double-ported valve, 179
Piston packing, 275
Piston rings, fitting, 271
Piston rod, 111
Piston valve, plain, 180
Plain D slide valve, 174
Plain eccentric valve gear, 193
  setting of, 194

Plain piston valve, 180
Pop safety valve, 12
  grinding, 94
Poppet type throttle, 122
Preparing boiler for storage, 66
  engine for storage, 108
Priming, boiler, 52
Pump, boiler feed: Clark, 24
  crosshead, 18
  geared, 22
  Marsh, 25
Pumps, oil, 131

Questions and answers, 286-307

Rear axle, 160
Reeves reverse gear, 219
  setting of, 220
Refitting, see Fitting
Regrinding, see Grinding
Repacking, see Packing
Repairing, boiler, tools for, 68
  friction clutch, 268
  governor, 264
  governor stem, 266
  leaky seams, 78
  rivet leaks, 86
  staybolt leaks, 76
Replacing staybolts, 74
Return flue boiler, 3
Reversing valve gear, 197
Rivet leaks, repairing, 86
Russel reverse valve gear, 222

Safety valve, 12
  grinding, 94
Scale, 54
Seams, repairing leaky, 78
Setting Arnold reverse gear, 216
  double eccentric reverse gear, 200

INDEX

Setting link reverse gear, 200
  Marsh reverse gear, 211
  Peerless reverse gear, 223
  plain eccentric, 194
  Reeves reverse gear, 220
  Russel reverse gear, 223
  Springer reverse gear, 209
  Stephenson reverse gear, 200
  Woolf reverse gear, 205
Shafts, counter, 159
  idler, 160
Single connection lubricator, 128
Slide, fitting, 234
Slide throttle, 123
Solid boxes, babbitting, 260
Springer reverse gear, 208
  setting of, 209
Starting new boiler, 42
  an engine, 149
Staybolts, replacing, 74
  repairing leaky, 76
Straw firing, 46
Steam chest, 104
Steam gauge, 10
  adjusting, 95
  testing, 95
Steering attachment, 162
Stephenson reverse gear, 198
  setting of, 200
Storage, of boiler, 66
  of engine, 168
Stop cock, grinding, 93

Table of average pressures, 178
  of position of eccentric and crank, 180
Tank, water, 165
Tender, 166
Testing boiler, 99
Testing steam gauge, 95
Throttle, butterfly, 122
  fitting, 94

Throttle locomotive, 123
  poppet, 123
  slide, 122
  wedge disk, 122
Tightening leaky tubes, 76
Tools, boiler repairing, 68
Traction engine, handling a, 169
Traction gearing, 145
Tubes, 49
  fitting new, 70
  tightening leaky, 76

Valve, Allen, 181
  Baker balanced, 192
  blow-off, 14
  compound engine, 188
  cut-off, fitting, 270
  direct, 178
  double-ported balanced, 182
  double-ported piston, 179
  gate, fitting, 94
  Giddings, 183
  Gould, 188
  indirect, 178
  Newton, 190
  Non-leak, 191
  plain D slide, 174
  plain piston, 180
  safety, 12
    grinding, 94
  Woolf, 185
Valve gear, reversing: Arnold, 215
  double eccentric, 198
  link, 198
  Marsh, 210
  Peerless, 223
  plain eccentric, 193
  Reeves, 219
  Russel, 223
  Springer, 208

INDEX

Valve gear, reversing: Stephenson, 198
  Woolf, 204
Valve gears, 192
Valve rods, packing, 275
Valve throttle: butterfly, 122
  fitting, 94
  locomotive, 123
  poppet, 123
  slide, 122
  wedge disk, 122
Valves, angle, grinding leaky, 91
Valves, check, grinding leaky, 91
  globe, grinding leaky, 91
Vertical boiler, 2

Water gauge glass, 8
Water, low, 62
Water tank, 165
Wedge disk throttle, 122
Wood, firing with, 45
Woolf reverse gear, 204
  setting of, 205
Woolf valve, 185

(1)